Mathematics in Industry

Volume 42

Mathematics in Industry focuses on the research and educational aspects of mathematics used in industry and other business enterprises. Books for *Mathematics in Industry* are in the following categories: research monographs, problem-oriented multi-author collections, textbooks with a problem-oriented approach, and conference proceedings. All manuscripts undergo rigorous peer review before acceptance. Relevance to the actual practical use of mathematics in the industry is the distinguishing feature of the books in the *Mathematics in Industry* series.

Philip J. Aston
Editor

More UK Success Stories in Industrial Mathematics

Springer

Editor
Philip J. Aston
Department of Mathematics
University of Surrey
Guildford, UK

ISSN 1612-3956 ISSN 2198-3283 (electronic)
Mathematics in Industry
ISBN 978-3-031-48682-1 ISBN 978-3-031-48683-8 (eBook)
https://doi.org/10.1007/978-3-031-48683-8

This Springer imprint is published by the registered company Springer Nature Switzerland AG
The registered company address is: Gewerbestrasse 11, 6330 Cham, Switzerland

Foreword

It will be apparent to anyone that the applications of mathematics have, for better or worse, been much in the news recently. Mathematical models lie at the heart of the methods that have been used both to predict the spread of COVID-19 and then to help decide what to do to reduce its impact. Mathematical models are also used to predict climate change, and to guide us in helping to prevent it. Mathematical models and algorithms based upon them, have also been used for many years by engineers and scientists, and they lie at the heart of much of modern technology. For example, your mobile phone would not work without the mathematical formulae coded within it, and a GPS system not only uses sophisticated mathematics, but also relies for its predictions on Einstein's General Theory of Relativity. We have also seen a recent, and significant, rise in the use of mathematically based methods and algorithms to guide decisions made by our policy makers. A good (or bad) example of this was the (mutant) algorithm used in 2020 to decide upon the results of the A-level examinations. But this is only part of the explosive growth that we are seeing in the use of machine learning and AI (which are heavily dependent on all areas of maths, including pure maths, applied maths, computational maths, and statistics) in nearly all areas of our life. Indeed, if you shop on Amazon, watch a movie on Netflix, order a taxi, or drive a car, then you will be using some aspect of machine learning.

Mathematicians have thus played a vital role in much of our modern society. But far from being praised for doing this, the use of mathematics in real world situations is often treated with suspicion. That is if it is considered at all. The use of mathematics in technology is often invisible. It is like the air we breathe. Without air we could not live, however, we cannot see it, and we often forget that it is there. The application of mathematics has also led to widespread concern, and suspicion, about the use of the predictions of mathematical models and algorithms to make decisions. But this is as much a problem of communication as of substance and it is important to set the record straight. Mathematical models are carefully constructed by taking well understood laws of physics, biology, or social science, formulating these as equations, and informing this whole process by using observational data. The equations are then solved (and this often involves a lot of sophisticated mathematics and computation) to give predictions. These predictions are then constantly tested against, and updated by, data taken from reality, and the model is refined, improved, and extended, to the point where its predictions can be trusted to represent reality. However, regardless of how carefully it is constructed, *a mathematical model is always going to be an approximation of reality*. Indeed, it is remarkable that an abstract subject such as mathematics is as good as it is in representing reality at all. (As the joke goes: to a physicist mathematics is a pale shadow of reality, to a pure mathematician, reality is a pale shadow of mathematics, but to a mathematical modeller, reality and mathematics are so intertwined as to be indistinguishable.) As soon as you involve people in the modelling process (as we had to do in the modelling of the response to COVID-19) then things get much more complicated and uncertain. It is important to note that whilst solving the equations is 'the fun bit' where mathematicians can really show what they can do, it is ALL of the above aspects of the modelling process which are important. All of the links are needed to make the chain work, and none are easier, harder, or better, than the others.

Despite all of the uncertainties, and difficulties, of applying mathematics to the real world as described above, using a mathematical approach is much the best way to help predict what might happen in the future. The advantage of using a mathematical model over informed guess work, is that you can be objective over the assumptions made in the model, including its powers and limitations. You can be quantitative in making predictions, which can then be tested. You can also run many 'what if?' scenarios to explore what might happen in different situations. Perhaps most importantly, although (as in all aspects of life) using a mathematical model involves a degree of uncertainty, you can quantify how much uncertainty there is in the model. Or to put it another way, you can say exactly how confident you are in its predictions. Modelling is about both solving the right problem, and also about solving the problem right. When dealing with the media it is important to communicate the fact that even if a mathematical prediction (for example for climate) is uncertain, it doesn't mean that it is wrong. But doing this well is not always easy!

During lockdown in 2020 (and of course during the run up to the Research Excellence Framework assessment REF 2021) I have been proud to have been part of the Virtual Forum for Knowledge Exchange in Mathematics (V-KEMS). V-KEMS was set up by a collaboration between the Isaac Newton Institute for Mathematical Sciences, the Newton Gateway, the International Centre for Mathematical Sciences, the Innovate UK KTN and a number of universities including my own. During lockdown, it ran very many online workshops which focused on the power of mathematics to bear on the challenges presented by COVID-19. These have ranged from keeping our shops, workplaces, schools, and universities safe from COVID-19, to helping the transport, healthcare, and leisure industries, recover from the devastating impact of COVID-19. These workshops were attended by mathematicians of all shapes and sizes, from pure mathematics, applied mathematics, statistics, and operational research. From young and old (with often Ph.D. students making major contributions), and from all over the UK and beyond. As well as V-KEMS, mathematicians were also heavily represented in SPI-M-O (Scientific Pandemic Influenza Group on Modelling, Operational sub-group) and on SAGE (Scientific Advisory

Group for Emergencies). It was a truly moving experience to see so many different mathematicians bring their skills together for the common good.

V-KEMS itself grew out of the rich tradition of 'study groups' which are hands on workshops for engaging mathematicians with industry. Originally started in Oxford in 1968, with an emphasis on applying differential equations to industrial problems, the study group concept is now in use all over the world. It is significant that recent study groups have become much more focused on data science, and machine learning and would seem much more 'pure' to an observer than they were even 20 years ago. Thus the rich tradition of all areas of UK mathematics in having a big impact on all aspects of life, which was celebrated in the first book on 'UK success stories in industrial mathematics' published in 2016, continues strongly to today, and the benefit of all.

I had a recent discussion with a journalist who said that whilst they agreed that mathematics was useful in understanding COVID-19, it was not 'real cutting edge mathematics'. I disagreed strongly, pointing out, for example, that ideas from stochastic partial differential equations and optimal transport were used to great effect, both of which were the subject of recent Fields Medals (widely regarded as the 'Nobel Prize of Mathematics').

So, to the subject of this book. The case studies presented in it, all derived from Impact Case Studies submitted to REF 2021, show much better than anything I can say, how vibrant the UK is in applying all areas of mathematics to make an impact to the real world. But even more important is the obvious fact, that to apply mathematics to any real situation, and to create impact, is hard, and you have to develop cutting edge mathematics to make any headway at all! It is a virtuous circle of great mathematics, leading to great applications, which lead to even greater mathematics, and so on. So I look forward even more to the next book, where we see the mathematics being created right now having an even bigger impact in the future.

December 2022

Chris Budd, OBE, FIMA, CMath
University of Bath
Bath, UK

Preface

"How do you do research in mathematics?" is a question I often get asked if I describe my job as an academic mathematician. I hope that this book will help to answer that question. Mathematics research contributes to the solution of problems in a multitude of disciplines with application to many facets of everyday life and society, but often goes unnoticed and unrecognised. So the intention of this book is to shine a light on some of the areas where mathematics and mathematicians are making fundamental contributions. The topics covered include climate and environment, engineering, health and healthcare, pharmaceuticals, animal welfare, business and finance, energy and food. The articles are wide ranging and include work done on:

- Improving weather forecasts
- Improving batteries for electric vehicles
- Real-time modelling during the COVID-19 pandemic
- Drug safety testing
- Digital marketing and social media networks
- A digital twin of a bank
- A collaboration between a pure mathematician and a composer.

In the UK, the Research Excellence Framework, or REF, takes place every 7 years or so which is an exercise to assess the quality of research being undertaken at all UK universities and in all disciplines. This includes a requirement to submit "Impact Case Studies", which describe situations where academic research has had an impact "beyond academia". All of these case studies are openly available on the REF website. However, they are written for a specific purpose and so are not easy reading for the generally interested reader. Thus, this book contains 36 articles, each describing one of these impact case studies that has been rewritten to tell the story of a successful collaboration between a mathematician and an end user in industry or government. All the articles describe the problem being considered, the mathematical solution, and the impact that the work has had. Our first book, *UK Success Stories in Industrial*

Mathematics, which was published by Springer in 2016, described some impact case studies that were submitted to REF 2014. This second volume describes more impact case studies that were submitted to the next REF in 2021.

This book will appeal to students and their lecturers as well as people working in industry, healthcare, finance, or government who have an interest in the use and application of mathematics. I trust that you will find these articles both interesting and informative, and then you may even be stimulated to explore the application of mathematics in your own sphere of interest.

Guildford, UK　　Philip J. Aston
August 2023

Contents

Engineering

Health

Healthcare

Energy

Food

Miscellaneous

Climate Modelling

Improving Met Office Weather and Climate Forecasts with Bespoke Multigrid Solvers

Andrew Malcolm, Eike Hermann Müller, and Robert Scheichl

Abstract At the heart of the Met Office climate and weather forecasting capabilities lies a sophisticated numerical model which solves the equations of large-scale atmospheric flow. Since this model uses semi-implicit time-stepping, it requires the repeated solution of a large sparse system of linear equations with hundreds of millions of unknowns. This is one of the computational bottlenecks of operational forecasts and efficient numerical algorithms are crucial to ensure optimal performance. We developed and implemented a bespoke multigrid solver to address this challenge. Our solver reduces the time for solving the linear system by a factor two, compared to the previously used BiCGStab method. This leads to significant improvements of overall model performance: global forecasts can be produced 10–15% faster. Multigrid also avoids stagnating convergence of the iterative scheme in single precision. By allowing better utilisation of computational resources, our work has led to estimated annual cost savings of £300k for the Met Office.

1 Introduction

Climate and weather forecasts have to be produced under very tight time constraints. For example, a typical model run to forecast the global weather for the next five days must complete within one hour, while predicting the evolution of several hundreds of millions of atmospheric variables. Efficient algorithms are crucial to achieve this goal and to make optimal use of computational resources on large, massively par-

A. Malcolm
Met Office, FitzRoy Road, Exeter EX1 3PB, UK
e-mail: andy.malcolm@metoffice.co.uk

E. H. Müller (✉)
Department of Mathematical Sciences, University of Bath, Bath BA2 7EX, UK
e-mail: e.mueller@bath.ac.uk

R. Scheichl
Institute for Applied Mathematics, University of Heidelberg, Im Neuenheimer Feld 205, 69120 Heidelberg, Germany
e-mail: r.scheichl@uni-heidelberg.de

P. J. Aston (ed.), *More UK Success Stories in Industrial Mathematics*,
Mathematics in Industry 42, https://doi.org/10.1007/978-3-031-48683-8_1

allel machines. The Met Office uses the semi-implicit dynamical core codenamed "ENDGame" [1] to predict large-scale flow. The time-stepping scheme requires the solution of a large, sparse system of linear equations in every model timestep, which can take up 25–60% of the total runtime. Due to the size of the problem, direct solver algorithms are infeasible and suitably preconditioned iterative methods have to be used. For many years the Met Office solver was based on a BiCGStab algorithm, preconditioned with a vertical line relaxation scheme, which takes into account the high aspect ratio of the grid. However, this solver suffers from large iteration counts. This becomes particularly troublesome if the solver is run in single precision: the Krylov subspace vectors can lose orthogonality and the solver might struggle to converge for tight error tolerances. To overcome this issue, we replaced BiCGStab with a bespoke multigrid solver. Three challenges had to be overcome:

1. After discretisation, the high aspect ratio of the grid results in an extremely anisotropic equation; this reduces the efficiency of standard multigrid algorithms with pointwise smoothers and uniform coarsening in all dimensions.
2. On a latitude-longitude grid, the gridlines converge at the poles, introducing an additional, spatially varying anisotropy in the horizontal direction.
3. Traditionally, parallel scalability of multigrid can be limited by the poor work-to-communication ratio on the coarse grids, reducing model performance.

To overcome the first challenge, we used the tensor-product multigrid approach proposed in [2]: the grid is only coarsened in the horizontal direction, and a block-Jacobi smoother takes care of the vertical anisotropy. The second challenge was addressed with conditional semi-coarsening [3]: close to the poles the grid is coarsened predominantly in the latitudinal direction, thus reducing the horizontal anisotropy on the coarser grids. To address the final issue, we exploit the fact that the linear equation correlates unknowns over approximately 10 grid points, independent of the model resolution and it is sufficient to use a "shallow" multigrid hierarchy with 3–4 levels.

The resulting multigrid method is robust and converges rapidly. Overall, a linear solve is about twice as fast as with the previously used BiCGStab solver.

2 Bespoke Multigrid Solvers

During each semi-implicit timestep the compressible Navier Stokes equations are solved with a quasi-Newton iteration. This in turn requires the solution of a linear system for the prognostic fields of (Exner-) pressure, wind speed, density and (potential) temperature. By eliminating all other atmospheric variables in a Schur-complement approach, a sparse linear system, akin to a discretisation of an elliptic PDE for the pressure correction, can be obtained [1]. This system has the form $Hu = b$, where H is a sparse, discrete Laplacian-like matrix, b is a given right-hand side and the vector u represents the pressure unknowns at gridpoints. Multigrid solvers are algorithmically optimal for this type of problem, in the sense that the each iteration reduces the residual norm $||b - Hu||$ by approximately one order of magnitude. Multigrid

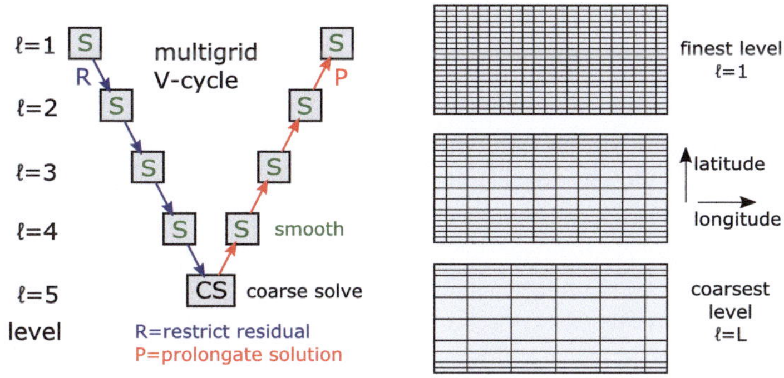

Fig. 1 Multigrid V-cycle (left) and conditional horizontal semi-coarsening (right)

constructs a series of coarser grids to speed up the solution of the equation. A small number of cheap smoother iterations is applied on the fine grid, which is followed by the solution of the restricted residual equation on a coarser grid, after which the coarse correction is prolongated back to the original fine grid where it is smoothed again. Applying the method recursively (Fig. 1), a much smaller equation is solved on the coarsest grid. The success of multigrid can be attributed to the reduction of the error at all length scales. However, the components of the algorithm, namely the smoother and coarse grid construction, have to be tuned to the problem at hand.

2.1 *Tensor Product Multigrid*

The elliptic operator $H = H_z + \delta H$ in ENDGame can be split into two components, where H_z contains the diagonal terms and couplings within a vertical column of the grid. Since these are much larger than the horizontal couplings collected in δH, the following stationary block-Jacobi method can be used to obtain a sequence of approximations $u^{(0)}, u^{(1)}, \ldots$ which converge to the true solution $u = H^{-1}b$:

$$u^{(k+1)} = u^{(k)} - \omega H_z^{-1}\left(b - Hu^{(k)}\right) \quad \text{for some overrelaxation factor } \omega \in \mathbb{R}. \quad (1)$$

While in the original ENDGame model the iteration in Eq. (1) is used to precondition BiCGStab, here we use it as a multigrid smoother. In [2] it was shown that if this approach is combined with horizontal-only coarsening of the grid, the resulting algorithm is equivalent to a 2d multigrid method. The convergence rate is determined by the *horizontal* Courant number, not by the much larger *vertical* Courant number.

2.2 Conditional Semi-coarsening

In addition to the vertical anisotropy, on a latitude-longitude grid there is also a spatially varying horizontal anisotropy $\alpha(\theta) = \Delta x_{\text{lon}}/\Delta x_{\text{lat}}$ which depends on latitude θ: close to the poles the latitudinal extent Δx_{lat} of grid cells is much larger than their longitudinal size Δx_{lon} and this makes the smoother less efficient. To address this issue, [3] propose to coarsen the grid uniformly in the longitudinal direction, but make the latitudinal coarsening conditional on the anisotropy $\alpha(\theta)$, thus coarsening more aggressively in this direction close to the equator. This reduces the anisotropy near the poles, which results in efficient smoothing on the coarser multigrid levels.

2.3 Shallow Multigrid

The convergence rate of the block-Jacobi iteration in Eq. (1) depends on the condition number of $H_z^{-1}H$. To a good approximation, this is given by the square of the horizontal Courant number $\nu = c\Delta t/\Delta x$ where c is the speed of sound. In the Met Office model $\nu \approx 10$, independent of the resolution. Crucially, the condition number is reduced by a factor 4 each time the grid is coarsened and Δx is doubled. This implies that after three multigrid coarsening steps the condition number is reduced to $\nu^2/4^3 \approx 1.56$. Hence, it is sufficient to limit the number of multigrid levels to 4 and use a simple iterative method to solve the problem on the coarsest grid.

2.4 Results

As Fig. 2 demonstrates, multigrid dramatically reduces the number of iterations and this leads to a significant acceleration of the solver. It also increases the robustness of the model, as demonstrated by the more monotonous convergence.

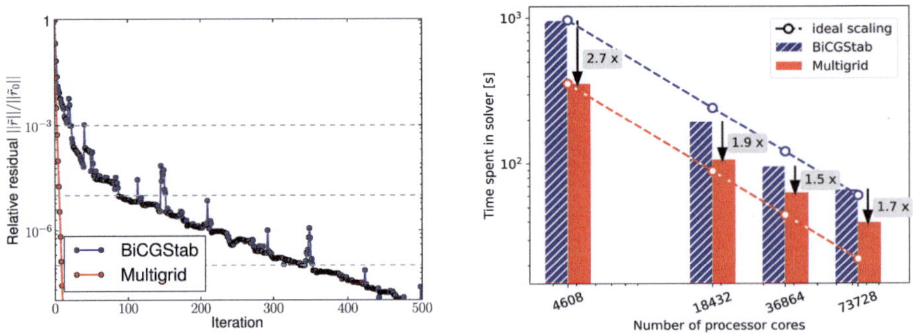

Fig. 2 Typical convergence history recorded early in the project (left) and parallel scaling of the time spent in solver for a 6 km operational global model run (right)

Table 1 Measured times for an operational global 10 km NWP run on a 2560 × 1920 × 70 grid

# cores	Total runtime (min)		Saving (%)	Solver time (min)		Saving (%)
	BiCGStab	Multigrid		BiCGStab	Multigrid	
4608	60.9	51.1	16	26.4	17.6	35
9216	32.3	28.2	13	14.3	10.2	29
11520	26.0	24.0	8	11.0	6.7	39

Table 1 shows the total runtime for a 75 h deterministic forecast in the current operational configuration of the model, which corresponds to a global computational grid with $2560 \times 1920 \times 70 = 3.4 \cdot 10^8$ points and a spatial resolution of around 10 km in mid-latitudes. In this case the total runtime is reduced by 8–16%, while accelerating the solver itself by 29–39%.

3 Achieved Impact

The key impact of this project consists in improvements to the Met Office numerical forecast products for climate and weather prediction. The code developed during the project has been used operationally since 9th December 2020. As explained in a recent news release (www.metoffice.gov.uk/research/news/2020/multigrid-solver):

> The new multigrid solver will allow higher resolution forecasts to be run in the future and is already allowing better utilization of supercomputer resources.

By convincing the Met Office to adopt novel multigrid technology, the research has impacted on important decisions at the Met Office and led to significant changes to computer codes which are crucial for the organisation. According to Dr Paul Selwood, a Principal Fellow in Supercomputing and Manager of the Met Office HPC Optimisation team:

> The overall investment of around 0.8 person-years (approximately £100 k) of senior staff into this project demonstrates the importance the Met Office is placing on [multigrid].

To demonstrate the significance of the achieved savings in model runtime, consider the resulting monetary savings through better supercomputer utilisation. The current yearly cost of all global operational runs is £3.7mio. As stated by Dr. Ben Shipway, Head of Dynamics Research at the Met Office:

> Operational tests [of multigrid] demonstrate an average reduction of 13% in total runtime (corresponding to significant cost savings of £300k per annum) [...] performance improvements of this order of magnitude are rare and typically only achieved once every 5 years.

The wider reach of the impact is demonstrated by the fact that the new multigrid code is currently used by one of the Met Office's international partners, the NIWA research institute in New Zealand.

3.1 Collaboration with the Met Office

Our research (see e.g. [4]) was carried out as part of the "GungHo" project. To convince the Met Office to pursue the new multigrid technology, the authors established a close working relationship with senior atmospheric scientists (Prof Nigel Wood, Dr Ben Shipway, Dr Thomas Melvin) and computational specialists (Dr Christopher Maynard, Dr Paul Selwood) through frequent mutual visits and joint meetings. To focus on the project, Eike Müller (EM) secured a secondment through the Bath Institute of Mathematical Innovation in 2017/18. The multigrid solver was documented and integrated into the ENDGame model by the authors (2012–2019), followed by rigorous internal code review and evaluation in operational configurations (2019–2020).

3.2 Future Work

Following the success of this project, the Met Office is committed to using multigrid technology also for its next generation model [5]. Here, multigrid has shown comparable speedups relative to Krylov subspace methods. EM has received further funding from EPSRC and the Met Office to explore the potential of multigrid for so-called hybridised discretisations, which are particular promising at higher order.

Acknowledgements This work was funded through NERC grants NE/K006754/1 and NE/J005576/1. The authors thank their collaborators at the Met Office and the members of the GungHo project. In particular, we would like to acknowledge the important contributions to the early stages of this project made by Dr Markus Gross, who tragically passed away in January 2022.

References

1. Nigel Wood, Staniforth, et al. An inherently mass-conserving semi-implicit semi-lagrangian discretization of the deep-atmosphere global non-hydrostatic equations. *Quarterly Journal of the Royal Meteorological Society*, 140(682):1505–1520, 2014.
2. Steffen Börm and Ralf Hiptmair. Analysis of tensor product multigrid. *Numerical Algorithms*, 26(3):219–234, 2001.
3. Sean Buckeridge and Robert Scheichl. Parallel geometric multigrid for global weather prediction. *Numerical Linear Algebra with Applications*, 17(2–3):325–342, 2010.
4. Eike H Müller and Robert Scheichl. Massively parallel solvers for elliptic partial differential equations in numerical weather and climate prediction. *Quarterly Joural of the Royal Meteorological Society*, 140(685):2608–2624, 2014.
5. Christopher Maynard, Thomas Melvin, and Eike Hermann Müller. Multigrid preconditioners for the mixed finite element dynamical core of the LFRic atmospheric model. *Quarterly Journal of the Royal Meteorological Society*, 146(733):3917–3936, 2020.

Improving Weather Forecasts Through Advanced Use of Observations

Sarah L. Dance and Nancy K. Nichols

Abstract Weather forecasts play a vital part in our lives, with major impacts on society and the economy. Forecasts are obtained by combining observations of the weather with computational predictions using a data assimilation process. Forecast accuracy relies on accurate estimates of the uncertainty in these weather observations. Remotely sensed observations, for example from satellites and ground-based instruments such as radar, provide the most benefit to forecast accuracy, but these are the most expensive data to acquire and their errors are not well understood. Research at the University of Reading has developed new methodology for estimating observation error statistics and has contributed important theoretical advances in understanding the role of observation uncertainty in weather forecasting. These methods are now being used in operational systems worldwide by the UK Met Office, European Centre for Medium-Range Weather Forecasts, US Naval Research Laboratory and NASA, amongst others. Thanks to these advances, there have been significant improvements in forecast accuracy and cost-effectiveness of observations for forecasting agencies, without loss of computational efficiency. The research has also influenced strategies for system enhancement, diagnostic activities and training at these centres and has raised the quality of weather information provided to government, business and the public for decision-making.

S. L. Dance
Department of Meteorology, University of Reading, Earley Gate, PO Box 243, Reading RG6 6BB, UK
e-mail: s.l.dance@reading.ac.uk

N. K. Nichols (✉)
Department of Mathematics and Statistics, University of Reading, Whiteknights, Reading RG6 6AX, UK
e-mail: n.k.nichols@reading.ac.uk

P. J. Aston (ed.), *More UK Success Stories in Industrial Mathematics*,
Mathematics in Industry 42, https://doi.org/10.1007/978-3-031-48683-8_2

1 Introduction

Weather forecasts are critical for transport, emergency services, agriculture, energy and defence sectors. National meteorological services typically use around ten million observations each day to improve global and regional predictions of weather on timescales from hours to weeks. Overall, improvement in operational weather forecast accuracy internationally increases value for money for forecasting agencies, who fund the data. Moreover, better forecast accuracy benefits society through protecting lives and mitigating economic losses as a result of adverse weather.

In weather forecasting systems, data assimilation is routinely used to improve predictions by combining forecasts from numerical model simulations with observations of the earth system in order to provide information on a wide range of variables such as temperature, pressures, winds and humidity. The assimilation problem is an enormous nonlinear Bayesian statistical estimation problem, where the model and observational data are weighted by their respective uncertainties. Forecast accuracy depends upon correctly specified uncertainties, as measured by error statistics. Better estimates of uncertainty result in better forecasts. A key research question is therefore how to characterise and treat observation uncertainty in an assimilation system.

Historically, the statistical relationships, or correlations, between errors in the observations have been difficult to estimate and so have been neglected in data assimilation. The operational norm in the past for the UK Met Office and other agencies around the world was to assume that the errors were uncorrelated. Researchers at the University of Reading explored, for the first time, the implications of the omission of observational error correlations in weather forecasts [1]. Defining 'information content' to be a measure of how much the use of observations reduces overall uncertainty in the assimilation system, the research indicated that neglecting these error correlations reduced the information content provided by these observations by 75%. The team then showed that inclusion in the assimilation of even a rough approximation to the true correlations can substantially increase information content and hence improve the weather forecast accuracy [2]. In tests with an idealized data assimilation system, the forecast errors could be reduced by a factor of three with little additional computation time.

To further increase information content and reduce uncertainty in forecasts, the Reading team worked with the UK Met Office to develop efficient mathematical methods for estimating observation error correlations. Ph.D. work in 2009, published in 2014 [3], showed for the first time that diagnostic methods that were originally designed as a consistency check could be used successfully to estimate the full observation error covariance matrix, including correlation terms, for operational data. The UK Met Office successfully applied these methods to estimate inter-channel observation error correlations for satellite radiance data in its system.

Researchers at Reading also made important theoretical advances in understanding observational errors and the effects of including correlated observation errors in the assimilation. Using spectral analysis, they investigated the sensitivity of the assimilation system and bounded differences in the estimated error variances and correlations due to changes in the system [4]. High sensitivity, measured by the condition number of the problem, was found by the UK Met Office to cause computational failures. In [5], Reading researchers mathematically analysed the problem, deriving upper and lower bounds on the condition number and showing that high sensitivity is related to complicated interactions between the error covariances of the observational and model data. Subsequent research focussed on improving the condition number of the system and reducing the computational time required.

Following the early work in [3], the team adapted their methods to quantify (i) inter-channel error correlations for a range of satellite instruments, and (ii) spatial error correlations, for example from satellites tracking clouds and from Doppler radar tracking precipitation and winds. The application to Doppler radar winds [4] by the Reading team led to international inter-comparisons between the UK Met Office and the German Weather Service (Deutscher Wetterdienst) assimilation systems. Through this research, issues affecting the assimilation of the observational data were identified and treated. These included biases in the data, inconsistencies in boundary treatment, and quality control issues in handling the data.

Research at Reading on the characterisation and treatment of observation uncertainty in assimilation systems for weather prediction has made significant contributions to the field, both theoretically and in practice. It has resulted in: international recognition of the value of accounting fully for observational errors [1, 2]; the development and refinement of new methodology for estimating observation error correlations [3, 6]; and important mathematical advances in understanding the role of correlated observation errors in assimilation systems [4, 5].

2 Method

In variational data assimilation, the aim is to minimize an objective function that measures the distance between the observations and their predictions from a known background state, given a description of their dynamic behaviour. The solution to the problem is known as the 'analysis'. The technique we use to estimate the observation covariances is based on linear estimation theory, where the optimal analysis is a Bayesian estimate weighted by the background and observation error covariances $\mathbf{B} = E[\epsilon^b(\epsilon^b)^T] \in \mathbb{R}^{n\times n}$ and $\mathbf{R} = E[\epsilon^o(\epsilon^o)^T] \in \mathbb{R}^{m\times m}$. Here E is the expectation operator, and we assume that the observation and background errors, $\epsilon^b \in \mathbb{R}^n$ and $\epsilon^o \in \mathbb{R}^m$, are unbiased and uncorrelated. The analysis, or estimated state of the atmosphere, $x^a \in \mathbb{R}^n$, is then expressed in terms of the known background state, $x^b \in \mathbb{R}^n$, the observation vector, $y \in \mathbb{R}^m$, and the background residual vector, $d^{ob} = y - H(x^b) \in \mathbb{R}^m$, where $H : \mathbb{R}^n \to \mathbb{R}^m$ is the observation operator mapping measurement space to state space. The equation is given by

$$x^a = x^b + \mathbf{BH^T}(\mathbf{HBH^T} + \mathbf{R})^{-1} d^{ob},$$

where $\mathbf{H}$ is the linearized observation operator. The analysis residuals are defined as $d^{oa} = y - H(x^a)$.

By taking the expectation of the cross product of the analysis and background residuals, we find a statistical approximation for the observation error covariance

$$E[d^{oa}(d^{ob})^T] = \bar{\mathbf{R}}(\mathbf{H}\bar{\mathbf{B}}\mathbf{H}^T + \bar{\mathbf{R}})^{-1}(\mathbf{H}\mathbf{B}\mathbf{H}^T + \mathbf{R}) = \mathbf{R}^e,$$

where $\mathbf{R}^e$ is the estimated covariance matrix and $\bar{\mathbf{R}}$ and $\bar{\mathbf{B}}$ are the estimates used in the assimilation. If the observation and forecast errors are exact, $\bar{\mathbf{R}} = \mathbf{R}$ and $\bar{\mathbf{B}} = \mathbf{B}$, then $E[d^{oa}(d^{ob})^T] = \mathbf{R}$. Due to the asymmetric use of the residuals, the approximation may not be symmetric and therefore we use the symmetric component $(\mathbf{R}_e + \mathbf{R}_e^T)/2$ of the diagnosed matrix. Studies have shown that even if the matrices $\bar{\mathbf{R}}$ and $\bar{\mathbf{B}}$ used in the assimilation are not specified exactly, the result still provides an approximation to the true matrix $\mathbf{R}$ that is better than assuming there are no correlations [2].

To demonstrate this technique we apply it to data from the IASI (Infrared Atmospheric Sounding Interferometer) instrument on board the satellite MetOp-A. IASI observations are an important component of the global observing system, with a positive impact on forecasting. The aim is to find the covariances between 139 of the observed IASI channels as shown in Fig. 1. The different colours represent the different groups of channels. For all channels we find that the diagnosed variances are considerably less than those that have been used in the past [3]. Traditionally it is assumed there are no correlations between these channels; however, we determine the correlation matrix to be as shown in Fig. 2. We find strong correlations between the channels in the water vapour (WV) and window channel bands. These are seen as diagonal red blocks in the figure. The full covariance matrix has now been applied to data assimilation in practice, leading to improvements in weather forecasting.

3 Impact

For the European Union alone, the economic benefits of weather forecasts are estimated at between EUR15bn and EUR61bn per year.[1] Because of the unknown error characteristics of the observational data, weather centres have previously found it difficult to achieve full value from expensively acquired satellite and radar observations, which motivated the research described in Sect. 2. The methods developed for estimating observation error statistics have enabled significant improvements in forecast accuracy. For example, the new method for dealing with correlated interchannel observation errors for the satellite instrument IASI became part of the UK Met Office operational global data assimilation system in January 2013 and has remained active since then. This resulted in a general improvement in forecast accuracy, with the most significant improvements being between 0.3 and 2.5% reductions

[1] https://www-cdn.eumetsat.int/files/2020-04/pdf_report_eps-sg_cost-benefit.pdf.

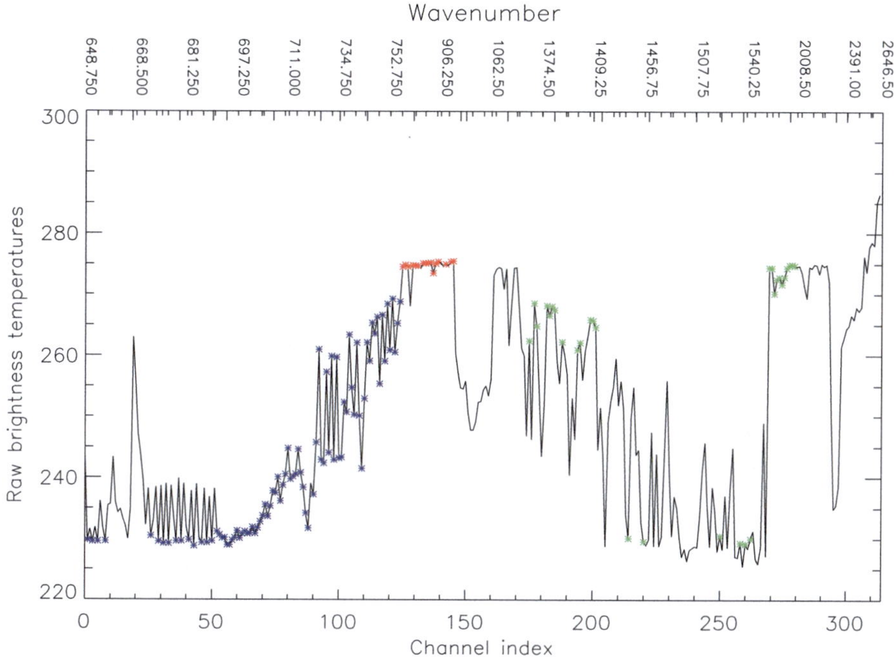

Fig. 1 Channels used in a variational scheme on a typical IASI spectrum (K): Temperature sounding channels indexed 0-86 (blue), window channels indexed 87–108 (red), and water vapour channels indexed 109–138 (green). Figure reproduced from [3], licensed under CC-BY 3.0. https://creativecommons.org/licenses/by/3.0/

in forecast root mean square error in key atmospheric variables. To put this in perspective, an improvement of around 3% is expected from a full year's worth of research and development improvements by the whole Met Office Weather Science Team of approximately 200 scientists. The method has been incorporated into operational weather prediction systems at major international weather centres around the world including the European Centre for Medium-Range Weather Forecasts (ECMWF), US National Centers for Environmental Prediction (NCEP), US National Aeronautics and Space Administration (NASA), US Naval Research Laboratory, Deutscher Wetterdienst (DWD), and RIKEN Centre for Computational Science (R-CCS), a Japanese national laboratory, with similar improvements in forecast accuracy.

The evidence provided by [4, 6] prompted the Met Office to re-parallelise its supercomputing software to enable spatial error correlations also to be handled. This has made it feasible, for example, to increase the number of weather radar wind observations assimilated by a factor of four, with no significant impacts on computation time. The new software has been used in the Met Office operational system since November 2019. Forecasts made with this system feed directly into weather warnings issued by the UK Public Weather Service and hence it is essential

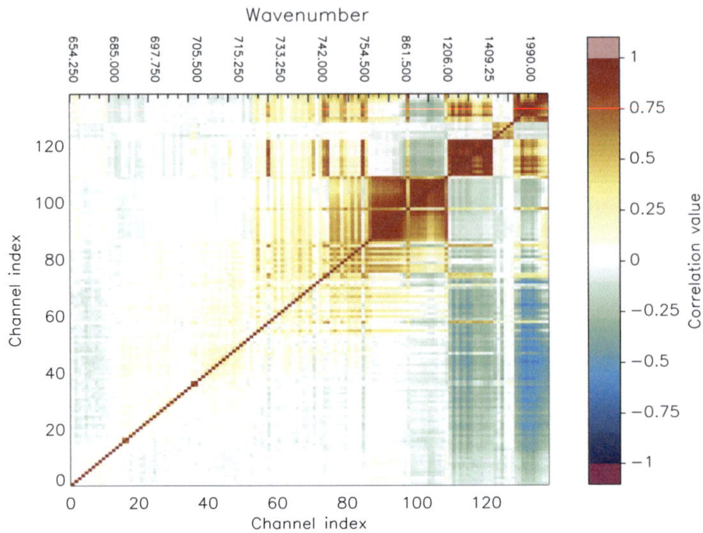

Fig. 2 Diagnosed observation error correlation matrix for the 139 channels used in a variational scheme. The wavenumber of the channels is given on the top axis and the colour bar indicates correlation value. Figure reproduced from [3], Figure 8, licensed under CC-BY 3.0. https://creativecommons.org/licenses/by/3.0/

that the computations be fast enough to ensure there is no delay in the provision of forecast products to customers and the public.

In addition, since 2016, the role of observation error correlations has become an integral topic in ECMWF's annual training course activities that are aimed at international, early career scientists working in operational weather centres and sponsored by EUMETSAT—a European intergovernmental organisation supplying satellite data, images and products to national meteorological services.

4 Conclusions

Weather services are of benefit across all sectors of society—from farming to flying. The value of a weather forecasting system depends on its accuracy, which is limited by the uncertainties in observations. Research at Reading has investigated the effects of observational errors on weather prediction, contributed to theoretical advances in the assimilation of the data and provided new methodology for better estimation of observation error correlations. The application of this research in major meteorological centres worldwide has proven that incorporating the observation error estimates in weather prediction systems increases forecasting accuracy and improves the cost-effectiveness of expensive data. Research at Reading has influenced strategic developments, diagnostic activities, and capacity building through training. With the rapid increase in the volume and types of satellite and radar data now available, the research continues to have a far-reaching influence on the best use of weather observations across the globe.

Acknowledgements Sarah L Dance and Nancy K Nichols were both supported in part by the NERC National Centre for Earth Observation. Sarah L Dance was also supported by an EPSRC Senior Fellowship EP/P002331/1.

References

1. Laura M Stewart, Sarah L Dance, and Nancy K Nichols. Correlated observation errors in data assimilation. *Int. J. Numer. Meth. Fl.*, 56(8):1521–1527, 2008.
2. Laura M Stewart, Sarah L Dance, and Nancy K Nichols. Data assimilation with correlated observation errors: experiments with a 1-D shallow water model. *Tellus A*, 65(1):19546, 2013.
3. Laura M Stewart, Sarah L Dance, Nancy K Nichols, John R Eyre, and J Cameron. Estimating interchannel observation-error correlations for IASI radiance data in the Met Office system. *Q. J. Roy. Meteor. Soc.*, 140(681):1236–1244, 2014.
4. Joanne A Waller, Sarah L Dance, and Nancy K Nichols. Theoretical insight into diagnosing observation error correlations using observation-minus-background and observation-minus-analysis statistics. *Q. J. Roy. Meteor. Soc.*, 142(694):418–431, 2016.
5. Jemima M Tabeart, Sarah L Dance, Stephen A Haben, Amos S Lawless, Nancy K Nichols, and Joanne A Waller. The conditioning of least-squares problems in variational data assimilation. *Numer. Linear Algebr.*, 25(5):e2165, 2018.
6. Joanne A Waller, David Simonin, Sarah L Dance, Nancy K Nichols, and Susan P Ballard. Diagnosing observation error correlations for Doppler radar radial winds in the Met Office UKV model using observation-minus-background and observation-minus-analysis statistics. *Mon. Weather Rev.*, 144(10):3533–3551, 2016.

Uncertainty Quantification for Climate Change

Michael Goldstein

Abstract The Met Office is tasked with developing a detailed uncertainty analysis for future climate, as part of a UK-wide climate change risk assessment, in order to meet the requirements of the Climate Change Act, 2008, which constructed a legally-binding long-term framework for building the UK's ability to adapt to a changing climate. This analysis exploited a general methodology for the Bayesian analysis of uncertainty for complex physical systems modelled by computer simulators. This report describes the general mathematical structure, its role in climate uncertainty assessment and the impacts, both statutory and in applications, arising from such climate projections.

1 Introduction

The Climate Change Act, 2008, constructed a legally-binding long-term framework for building the UK's ability to adapt to a changing climate. The Act requires a UK-wide climate change risk assessment, setting out the Government's policies for responding to the risks so identified by means of a National Adaptation Programme. This programme draws heavily on the uncertainty analysis for future climate outcomes carried out by the Met Office. The resulting climate projections further impact on all industries and public sector organisations who are likely to be affected by climate change and who therefore make decisions on policy and investment which must address the uncertainties in future climate.

The analysis carried out by the Met Office exploited a general methodology for the Bayesian analysis of uncertainty for complex physical systems modelled by computer simulators. There is an extensive literature on this area, including, in particular, fundamental research carried out at Durham University, which underpinned important aspects of the Met Office analysis. In this article, we describe this general

M. Goldstein (✉)
Department of Mathematical Sciences, Durham University, Durham DH1 3LE, UK
e-mail: michael.goldstein@durham.ac.uk

P. J. Aston (ed.), *More UK Success Stories in Industrial Mathematics*,
Mathematics in Industry 42, https://doi.org/10.1007/978-3-031-48683-8_3

mathematical structure, its relevance to climate projections and the impact that such projections have both in the framework of statutory regulation and in public actions relevant to anticipating the effects of climate change.

2 Mathematical Structure for Uncertainty Quantification

Much work has been done on the general subject of Bayesian analysis of uncertainty for complex real world physical systems modelled by computer simulators. The simulator may be represented as a function, f, which takes as inputs a vector x representing properties of the system, plus various boundary conditions and forcing functions, and returns as outputs a vector $y = f(x)$ representing aspects of the behaviour of the system. For example, in a climate system, y might be a vector of climate outputs, such as global temperature profiles, while the inputs might be parameters of the equations which determine how global system properties, such as CO2 levels, transform into such temperature profiles. The simulator will not give a precise representation of the physical system, the difference being referred to as the structural discrepancy of the simulator. The simplest representation of such discrepancy is to suppose that $y = f(x^*) + \epsilon^*$, where x^* is an appropriate choice for x and ϵ^* is a random vector expressing the difference between the simulator and the system for this choice. There are many different versions of this structure depending partly on whether we view there to be one, many or possibly no appropriate choices for x^* and partly on the error structure that we impose on ϵ^*, for example whether we consider it to be independent of $f(x^*)$. Often, we have observations, z, with noise, on a subvector y_h of y. For example, we may suppose that $z = y_h + \epsilon_h$, where ϵ_h is a random noise term independent of y_h.

Typical aims for our uncertainty analysis are to make an inference about x^* given z and to exploit that inference to improve our forecasts for those elements of y which have not yet been observed. The basic complications of this analysis arise from the high dimensional nature of the input and output spaces, the complexity of the function f and of the structural discrepancy process and the length of time that it takes to make an evaluation of $f(x)$ for a single choice of x. Fast surrogate forms, termed emulators, which approximate the computer model, with an associated uncertainty assessment of the quality of the approximation, are integral to this approach, both to address issues of evaluation speed for slow and expensive models and also to bridge important gaps arising from model imperfections.

In particular, research carried out by the statistics group at Durham University concerns the development of a very general probabilistic framework for linking mathematical models to the physical systems that the models purport to represent. This framework takes account of all sources of uncertainty, including model and simulator imperfections, and is a necessary precondition for making probabilistic statements about the system on the basis of historical observations and evaluations of the computer simulators. The formulation distinguishes simulators according to their quality and the nature of their inputs. Further modelling constructs are introduced to account for imperfections in the available simulators and to unify the composite inference, from the collection of available simulators, observed historical data and the judgements of experts, for the behaviour of the actual physical system. An early case study of our approach is contained in [1]. An exemplar of the general way that we may develop tractable forecasts for such complex systems is contained in [2]. The framework is general, and the structure is described in [3], but the ideas were developed with the application to collections of evaluations of climate models in mind, see [4]. Therefore, one aspect of this research is work which develops the underlying structure in such a way that it may be directly applied to relevant problems in climate science. This application of the research forms the core of the impact case described in this report. For a detailed overview of these issues in climate science, see [5].

3 Uncertainty for Climate Projections

The UK Climate Projections 2018 (UKCP18), an updated version of the UK Climate Projections 2009 (UKCP09), is the Met Office's climate analysis tool. It provides projections for UK and global climate change throughout the 21st century using the latest climate science and analysis. It is designed to assist decision-makers by enabling them to assess climate change risks and is used by both public and private users to plan for a wide spectrum of environmental impacts. UKCP09 provided authoritative climate forecasts which were widely used in climate adaptation planning, within a statutory legal framework that we will describe below, until they were replaced by the enhanced forecasts arising from UKCP18. The science and methodology used to construct UKCP09 is described in the report [6], from the Met office. The corresponding document for UKCP18 is [7].

Each report emphasises the importance of the careful treatment of uncertainty in the climate projections. Here is an indicative quotation. "Uncertainty in climate change projections is a major problem for those planning to adapt to a changing climate. Adapting to a smaller change than that which actually occurs (or one of the wrong sign) could result in costly impacts and endanger lives, yet adapting to too large a change (or, again, one of the wrong sign), could waste money. In addition there is the risk of maladaptation—adapting to climate change in a way that prevents or inhibits future adaptation. The 2008 projections are the first from UKCIP to be designed to treat uncertainties explicitly ... This means that probabilities are attached to different climate change outcomes, giving more information to planners and decision makers" [6, p. 19].

The methods developed at Durham play an important role in the uncertainty analysis throughout each report and Durham research is referred to explicitly in the UKCP09 and UKCP18 documents. Here are two indicatives quotes from the UKCP18 document, each citing Ref. [3].

"These results support the chosen approach, allowing Strand 1 to provide an updated product that combines evidence from HadCM3-basedPPEs and CMIP5 models in a consistent manner, retaining the Bayesian statistical framework used for UKCP09 (Goldstein and Rougier [3]). The probabilistic projections are provided for a core set of UK climate variables (Fung et al., 2018). The new projections remove the requirement previously placed on users to evaluate the probabilistic projections and CMIP5 results as independent lines of evidence" [7, p. 11]. "Prior pdfs ... are then produced using the Bayesian method of Sexton et al (2012), based on the general framework of Goldstein and Rougier (2004)" [7, p. 17].

The contribution of our approach for UKCP09 and UKCP18 is amplified in the above referenced paper [8], which explains in detail the uncertainty methodology used by UKCP09 and UKCP18. For example, Sect. 3 of the paper, "Outline of the calculations", begins "Here we describe the general steps in Goldstein and Rougier (2004) necessary to determine a probability distribution of some aspects of climate change that we want to predict" and Sect. 6.2 begins "... Goldstein and Rougier (2004) gives us several key advantages. ...First, the multivariate nature of this probabilistic framework allows us to have more than one prediction variable. Predicting joint probabilities provides us with important information on how uncertainty is related across different climate variables.....".

4 Impact: Statutory Climate Responsibilities

With the Climate Change Act (CCA) 2008, the UK constructed a legally-binding long-term framework to cut greenhouse gas emissions and a framework for building the UK's ability to adapt to a changing climate. The Act requires a UK-wide climate change risk assessment (CCRA) to take place every five years. The CCRA constructs sector reports describing a wide range of potential risks in each of the following sectors (followed by a more detailed analysis of selected risks that were judged to be the most important): Agriculture; Biodiversity & Ecosystem Services; Built Environment; Business, Industry & Services; Energy; Floods & Coastal Erosion; Forestry; Health; Marine & Fisheries; Transport; Water.

The CCRA is constructed to facilitate the Climate Change Act mandated National Adaptation Programme, (NAP, 2013) and updated version, in 2017, setting out the Government's objectives, proposals and policies for responding to the risks identified in the CCRA. The NAP (laid before Parliament in 2013) report explains: "The Climate Change Risk Assessment 2012 (CCRA) for the UK brought together the best available evidence, using a consistent framework to identify the risks and opportunities related to climate change. The assessment distilled approximately 700 potential risks down to more than 100 for detailed review. The government's response to the CCRA, which meets the requirements laid down in the Climate Change Act 2008, is the first NAP. In developing the NAP for England, we have taken the highest order risks from the CCRA and working in partnership with businesses, local government and other organisations, have developed objectives, policies and proposals to address them".

The original 2012 CCRA, and thus the NAP, draws heavily on the uncertainty analysis in UKCP09 to take account of the many uncertainties including natural climate variations, projected emissions, and the specific effect on UK climate of global warming. The impact of that report was still being felt heavily through the REF impact period of this study. The second report was published in 2017 and preparations for the third report are underway.

The original CCRA used the UK Climate Projections (UKCP09) for three time periods—30-year periods centred on the 2020s, 2050s and 2080s. The CCRA attempted to monetise the most important risks to the UK, and concluded that the results indicated that the net economic costs to the UK are of the order of tens of billions/year by the 2050s (in current prices) even for the middle of the uncertainty range for these costs at the middle of the projected emission scenarios. UKCP09 therefore played a key role within the Government's statutory responsibilities for assessing and responding to climate change throughout the period of this study, and the updated UKCP18 is continuing this role. The ministerial foreword to the updated version, NAP 17, links the various strands of activity as follows. " Later this year we will be launching a revised set of UK climate projections ('UKCP18'), replacing the current 2009 set and providing the most up to date and scientifically robust estimations of climate scenarios out to the end of this century. These projections are a key tool to enable everyone to future proof policies and activities to ensure our resilience to possible future climate change. ... so that we take back control of our fisheries and agriculture, restore nature and care for our land, rivers and seas. Using a natural capital approach ensures that we take account of all the many benefits our environment provides, and we will develop policies that ensure our seas and lands are healthy and productive and resilient to climate change".

The UK Climate Projections 2018 (UKCP18) were launched on 26 November at an event at the Science Museum in London with national media in attendance. The announcement highlighted the need for further urgent action. Opening the launch, Environment Secretary Michael Gove said:"This cutting-edge science opens our eyes to the extent of the challenge we face and shows us a future we want to avoid. By having this detailed picture of our changing climate, we can ensure we have the right infrastructure to cope with weather extremes, homes and businesses can adapt, and we can make decisions for the future accordingly".

5 Public and Private Sector Impact

In addition to the statutory role, the Met Office has worked with a wide range of public and private sector organisations to use UK climate impacts to inform decisions on investment amounting to billions of pounds to future proof projects against climate change.

On the (archived) UKCP09 website are described case studies showing how the UKCP09 products have been used. The range of topics covered includes: national assessment of river flows, climate change and pollution, strategic planning for flood management, changes in flood damages at a catchment scale, assessments of storm surge and sea level rise, emergency planning, defining land capability for agriculture specifications, assessing potential vulnerabilities to climate change, future proofing design decisions in the buildings sector, investigating coastal recession & shore profile development, storm surge and sea level rise and assessing impacts of climate change on tourism in South West England.

As evidence of the ways in which the Met Office are rolling out the effects of UKCP18 to user groups, the website https://www.metoffice.gov.uk/research/approach/collaboration/ukcp/using-ukcp/case-studies describes current case studies of uses of UKCP18 in areas such as flood risk and forestry management and also details a set of six project leaflets demonstrating how different sectors may use the new climate projections. The leaflets are based on a set of demonstration projects led by sector specialists and were written for their sector. Each leaflet describes how UKCP09 has been used in the area and how UKCP18 will improve adaptation planning for climate change. The titles show some of the range of areas of application that UKCP18 affects. (i) Assessing climate change risk in Yorkshire; (ii) Future surface water flood hazard risk; (iii) Coastal cliff recession under climate change; (iv) Thermal performance of buildings; (v) Forests for the future; (vi) Water resources and drought planning.

References

1. P.S. Craig, M. Goldstein, A.H. Seheult, J.A. Smith (1997), Pressure matching for hydrocarbon reservoirs: a case study in the use of Bayes linear strategies for large computer experiments (with discussion), in Case Studies in Bayesian Statistics, vol. III, eds. C. Gastonis et al. 37–93. Springer-Verlag. New York, I
2. P. Craig, M. Goldstein, J. Rougier, and A. Seheult, (2001) Bayesian forecasting for complex systems using computer simulators, J. Amer. Statist. Assoc., 96 (2001), pp. 717–729.
3. Goldstein, M and Rougier, J (2004) Probabilistic formulations for transferring inferences from mathematical models to physical systems, Siam J. Sci Comput., 26, 467–487
4. Rougier, J. C. (2007). Probabilistic inference for future climate using an ensemble of climate model evaluations. Climatic Change, 81, 247–264
5. Rougier JC and Goldstein M (2014) Climate Simulators and Climate Projections, Annual Review of Statistics and Its Application Vol. 1:103–123.
6. Murphy JM, Sexton DMH, Jenkins GJ, Booth BBB, Brown CC, Clark RT, Collins M, Harris GR, Kendon EJ, Betts RA, Brown SJ, Humphrey KA,, McCarthy MP, McDonald RE, Stephens A, Wallace C, Warren R, Wilby R, Wood RA (2009) UK climate projections science report: climate projections Met Office Hadley Centre, Exeter.
7. J.M. Murphy, G.R. Harris, D.M.H. Sexton, E.J. Kendon, P.E. Bett, R.T. Clark, K.E. Eagle, G. Fosser, F. Fung, J.A. Lowe, R.E. McDonald, R.N. McInnes, C.F. McSweeney, J.F.B. Mitchell, J.W. Rostron, H.E. Thornton, S. Tucker and K. Yamazaki (2018) UKCP18 Land Projections: Science Report.
8. Sexton DMH, Murphy JM, Collins M, Webb MJ (2012) Multivariate probabilistic projections using imperfect climate models part 1:outline of methodology, Climate Dynamics, 38, 2513–2542

Enhancing Computational Models for Weather and Climate Forecasting

James Kent

Abstract Assessing atmospheric models is essential to ensure that these models produce accurate weather and climate forecasts. The dynamical core, the fluid dynamics component of the model, is the building block of the model, and it, along with its ability to be coupled with physical parametrizations, needs to be evaluated systematically. This chapter outlines the author's contributions to the Dynamical Core Model Intercomparison Project (DCMIP), which developed a series of dynamical core test cases that have now become a standard in the community, and his contribution to developing the transport component of NASA's GEOS-5 linear model.

1 Introduction

Accurate and reliable weather forecasts and climate predictions depend upon the quality of the computational models that simulate the evolution of the atmosphere. Global climate modelling systems consist of many different components, such as atmospheric, oceanic, land and sea ice models. The atmospheric model comprises two key parts: the dynamical core, responsible for capturing the atmospheric dynamics via numerical integration of the governing equations, and the physical parameterizations, which provide an approximate representation of unresolved processes such as small-scale turbulence. Significant mathematics is needed to design the dynamical core, such as the choice of equation sets, the numerical methods used to solve the equations, the geometry used to grid the sphere, and computational performance aspects. Some atmospheric models employ a third part: the linear model, a linearized version of the dynamical core that is used in data assimilation and sensitivity studies.

This work was carried out whilst the author worked at the University of South Wales, Pontypridd, UK.

J. Kent (✉)
Met Office, FitzRoy Road, Exeter, Devon EX1 3PB, UK
e-mail: james.kent@metoffice.gov.uk

P. J. Aston (ed.), *More UK Success Stories in Industrial Mathematics*,
Mathematics in Industry 42, https://doi.org/10.1007/978-3-031-48683-8_4

Operational weather and climate centres often develop their own atmospheric dynamical cores. It is important to assess the dynamical core in isolation, to understand whether the numerical methods used are capturing the required phenomena (such as correct wave propagation, or precipitation when dynamics and physics are coupled). A standard set of idealized test cases, with known solutions or features of solutions, are required to address this need. These test cases can then be used by operational centres to evaluate the performance of their model whilst in development, and hence influence key design aspects of their dynamical cores leading to model improvements. The dynamical core modeling community can then systematically compare the strengths and weaknesses of different model designs.

This article summarizes the author's contributions to the Dynamical Core Model Intercomparison Project (DCMIP), and his work with NASA's linear model.

2 Dynamical Core Testing

DCMIP was set up to establish a community that compares developing and operational dynamical cores, and to provide a forum to exchange ideas and advance education on model development. Although test cases exist for simplified equation sets (see, for example, [1]), the aim of DCMIP was to develop a standard test suite for dynamical cores.

The governing equations for an atmospheric dynamical core can be written in many different ways. One such example is given below:

$$\frac{\partial \mathbf{u}}{\partial t} + (\mathbf{u} \cdot \nabla)\mathbf{u} + 2\Omega \times \mathbf{u} + \nabla\Phi + c_p\theta\nabla\Pi = \mathbf{S}_u, \quad (1)$$

$$\frac{\partial \theta}{\partial t} + \mathbf{u} \cdot \nabla\theta = S_\theta, \quad (2)$$

$$\frac{\partial \rho}{\partial t} + \nabla \cdot (\mathbf{u}\rho) = 0, \quad (3)$$

$$\Pi^{(1-\kappa)/\kappa} + \frac{R}{p_0}\rho\theta = 0. \quad (4)$$

Here $\mathbf{u}$ is the velocity vector, t is time, Ω is the rotation rate vector, Φ is the geopotential, θ is the potential temperature, c_p is the specific heat capacity at constant pressure, R is the gas constant for air, κ is the ratio of R to c_p, $\Pi = (p/p_0)^\kappa$ is the Exner pressure, ρ is the density, p_0 is a reference pressure, and S indicates a source term.

The governing equations include the momentum equations (1), a thermodynamic equation (2), the continuity equation (3), and the equation of state (4). The source terms are from the physical parameterizations. This is a complex system of 5 prognostic partial differential equations and 1 diagnostic equation. In addition to these dynamic equations there are transport equations for moisture and gas species. This whole equation set can be broken down to facilitate simplified testing of different components, e.g. just the transport terms [2]. The 2012 DCMIP workshop focused on the dry dynamics without physics, i.e. with $S = 0$. The next stage, and a gap in the literature, was a set of test cases to investigate moist dynamics and the coupling of the dynamics with the parameterized physics.

For DCMIP 2016 the author helped design and develop the test cases to achieve this. A *simple* physics package was developed to provide the physics forcings in an idealized way, and a framework was developed to set up the tests. This framework included shrinking the radius of the Earth so that the vertical and horizontal length scales are similar. The three test cases were:

- Moist baroclinic instability test with chemistry;
- Moist tropical cyclone test;
- Mesoscale supercell storm test.

The tests were designed to have relatively straightforward model initialisation, yet provide challenging environments for the dynamical cores. The 2016 DCMIP workshop invited modelling centres to evaluate their dynamical cores using these tests [3].

3 Improving the Linear Model

The DCMIP test cases focused on dynamical cores and simple coupling to physics. Related to the dynamical core is the linear model, which is used to describe the forward evolution of perturbations and the backwards evolutions of sensitivities. These are required to measure the discrepancies in the model with respect to observations, and due to the linear models computational efficiency, it is a key tool in data assimilation and the estimation of sensitivity to initial conditions.

During discussions at DCMIP the author became aware that although the different components of an atmospheric model are coupled together, often they are developed independently, with design choices being chosen to suit one component. This was the case with NASA's GEOS-5 dynamical core and associated linear model. The optimum numerical methods for the full dynamical core are not always the optimum methods for the linear model.

NASA's GEOS-5 model contains a linear model which is based on a linearized version of the full model. However, much of the full model is nonlinear, and thus the linearization does not always produce the optimal results. One aspect of this is the tracer transport algorithm. Many quantities (such as density, temperature, gases, even the winds themselves) are transported throughout the atmosphere by the winds. The transport algorithm is the part of the system that models this.

The transport equation for a gas (called a tracer), denoted q, is the advection equation:

$$\frac{\partial q}{\partial t} + \mathbf{u} \cdot \nabla q = S, \tag{5}$$

where $\mathbf{u}$ is the three-dimensional velocity vector, t is time, and S is any source term. For a simplified case let's consider a one-dimensional problem. Consider q transported by a constant velocity $u > 0$ in a periodic domain x with no source term. This can be represented using the one-dimensional advection equation:

$$\frac{\partial q}{\partial t} + u\frac{\partial q}{\partial x} = 0. \tag{6}$$

The tracer q is just translated across the domain and a true solution is known.

This equation is linear in q and if solved using a linear scheme it can be used directly in the linear model. However, it is common for atmospheric dynamical cores to use non-linear transport schemes that include switches called limiters.

Desired properties of a transport scheme include high-order accuracy and monotonicity. A limiter switches between a high-order scheme and a low-order monotonic scheme to achieve high-order accuracy when data is smooth, but prevent oscillations when the data is not smooth. For example, a reconstruction of q, denoted $\tilde{q}$, used to compute the $\partial q/\partial x$ term would have the form

$$\tilde{q} = \tilde{q}_{\mathrm{L}} + \phi\left(\tilde{q}_{\mathrm{H}} - \tilde{q}_{\mathrm{L}}\right), \tag{7}$$

where subscript L indicates a low-order reconstruction, subscript H indicates a high-order reconstruction, and ϕ is the limiter. An example of a limiter is the well known van Leer limiter which has the form

$$\phi = \frac{r + |r|}{1 + |r|} \tag{8}$$

where r is a smoothness parameter

$$r = \frac{q_j - q_{j-1}}{q_{j+1} - q_j} \tag{9}$$

and j is the spatial index. As ϕ is a function of q, usually involving gradients to assess smoothness, this scheme is non-linear in q. In fact, when $r = 0$ the derivative of ϕ is discontinuous and therefore no exact linearization exists. Developing a linear scheme based on this type of scheme becomes much harder, and in fact can lead to errors as the linearized model is not not necessarily accurate in the limit of small perturbations.

These transport schemes with limiters are necessary for the dynamical core (see Fig. 1) but are an issue for the linear model. We investigated this issue first using simple one-dimensional advection problems [4]. Consider a variable q linearized into reference (superscript r) and perturbation (prime) parts, $q = q^r + q'$. We systematically compared the results from the linearized scheme for q' with the results of advecting perturbed and unperturbed fields (q and q^r respectively) using the full non-linear scheme to show how correlated the linearization is with the full scheme.

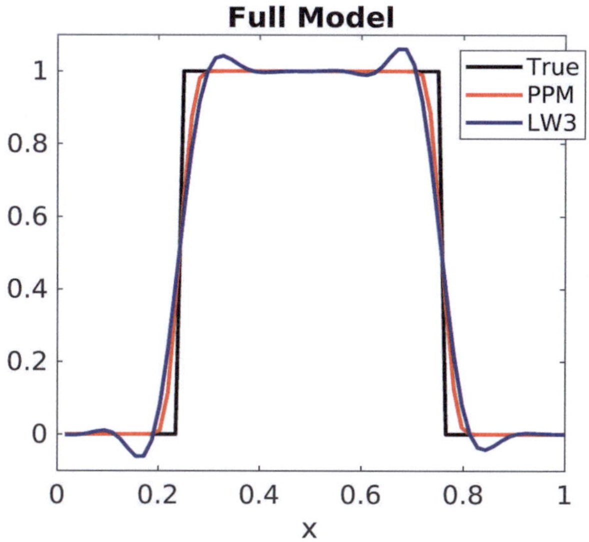

Fig. 1 Transport of a step function (black) once around the domain using the PPM scheme (red) and the 3rd order Lax–Wendroff scheme (blue). The PPM scheme uses a limiter and so maintains the shape of the step, as well as the maximum and minimum values, well. The Lax–Wendroff scheme is linear and has produced unphysical over- and undershoots

A key finding was that although monotonicity is needed for transport in the dynamical core, it is not a vital part of the linear model's transport. This means that the linear model transport scheme doesn't have to be an linearization of the non-linear transport scheme. The results from [4] showed that using a separate linear scheme for the linear model transport produced better correlations than using an attempted linearization of the full scheme when a non-linear scheme was used (see Fig. 2).

In GEOS-5 the transport scheme is based on the piecewise parabolic method (PPM), which fits a parabola within a cell and then integrates to find the mass passing through the flux point. Limiters are used to ensure monotonicity. The linearization of PPM becomes very complex, and its use in the linear model meant that the GEOS-5 linear model was not well correlated with the full dynamical core. To overcome this problem we used a linear scheme that had similar properties to PPM, such as order-of-accuracy, but was not just a linearization of PPM. A third-order Lax–Wendroff type scheme was implemented within the GEOS-5 linear model. Testing shows that this new method improved the correlations in the idealised testing and within GEOS-5 [4].

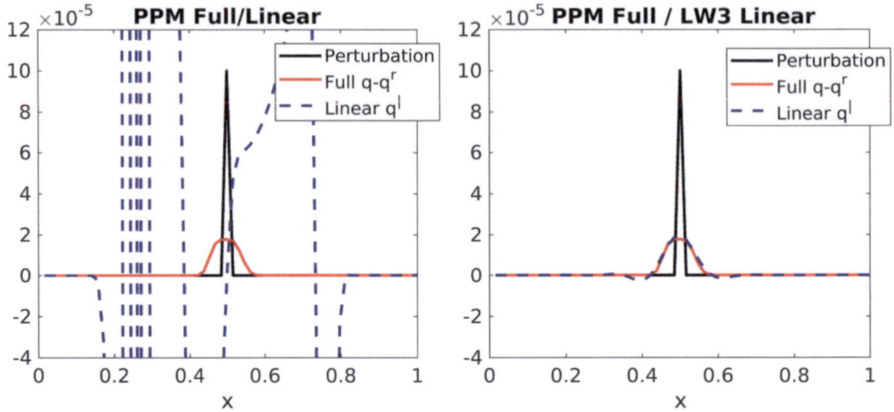

Fig. 2 Comparing the full scheme $q - q^r$ solution (red) with the linearized scheme q^l solution (blue) transported once around the domain. If they are well correlated the red and blue lines should coincide. The left plot shows the results when using the PPM scheme as the full scheme and using the linearization of PPM for the perturbation. The right plot shows the results when using the PPM scheme as the full scheme and using the 3rd order Lax–Wendroff scheme for the perturbation. Using the linearization of PPM produces very poor results. Using Lax–Wendroff for the perturbation greatly improves the correlation

4 Details of Impact

Accurate weather and climate predictions depend upon accurate numerical methods solving the governing equations. The development of these methods, along with their careful assessment and evaluation, is essential for improving forecasting capabilities and having confidence in the model results.

The test cases developed for the 2016 DCMIP workshop have become a standard test bed for developing and assessing atmospheric dynamical cores. These tests allow modelling groups to assess complex interactions between dynamics and physics with quick turn-around times, and to compare results with other groups in the community [5]. These tests were used by the modelling groups that attended the workshop. These groups included major national and international forecasting centres, such as:

- National Center for Atmospheric Research (NCAR), Boulder, CO, USA.
- Geophysical Fluid Dynamics Laboratory (GFDL), Princeton, NJ, USA.
- European Centre for Medium-Range Weather Forecasts (ECMWF), Reading, UK.
- Environment and Climate Change Canada (ECCC), Dorval, Québec, Canada.
- Deutscher Wetterdienst (DWD), Offenbach am Main, Germany.
- Naval Research Laboratory, Monterey, CA, USA.
- Japan Agency for Marine-Earth Science and Technology, Yokohama, Kanagawa, Japan.

During the workshop bugs were found and corrected in developers model code. The library of results gives developers the confidence that their model is behaving in the correct manner, and can highlight potential deficiencies and limitations of their dynamical cores. The set of test cases developed during DCMIP is still widely used in model development.

Our method for the linear model transport has become the default option in NASA's GEOS-5 linear model. Using our approach has improved the accuracy, stability and computational efficiency of the linear model [4]. The GEOS linear model is being used by the Joint Center for Satellite Data Assimilation (JCSDA) to develop the JEDI data assimilation system.

Acknowledgements Thanks must go to Dan Holdaway (NOAA/NWS/NCEP/EMC) for collaborating on the NASA GEOS-5 linear model, and to Christiane Jablonowski (University of Michigan), Paul Ullrich (UC Davis), Kevin Reed (SUNY), Colin Zarzycki (Penn State), Peter Lauritzen (NCAR), Ram Nair (NCAR) for our work together on the DCMIP workshops.

References

1. Williamson, D. L., Drake, J. B., Hack, J. J., Jakob, R. and Swarztrauber, P. N.: A Standard Test Set for Numerical Approximations to the Shallow Water Equations in Spherical Geometry. J. Comput. Phys., **102**, 211–224. (1992)
2. Kent, J., Ullrich, P. A., and Jablonowski, C.: Dynamical Core Model Intercomparison Project: Tracer Transport Test Cases. Q. J. Royal. Meteorol. Soc., **140**, 1279–1293. (2014)
3. Ullrich, P. A., Jablonowski, C., Kent, J., Lauritzen, P. H., Nair, R., Reed, K. A., Zarzycki, C. M., Hall, D. M., Dazlich, D., Heikes, R., Konor, C., Randall, D., Dubos, T., Meurdesoif, Y., Chen, X., Harris, L., Kuhnlein, C., Lee, V., Qaddouri, A., Girard, C., Giorgetta, M., Reinert, D., Klemp, J., Park, S.-H., Skamarock, W., Miura, H., Ohno, T., Yoshida, R., Walko, R., Reinecke, A., and Viner, K.: DCMIP2016: A Review of Non-Hydrostatic Dynamical Core Design and Intercomparison of Participating Models. Geosci. Model Dev., **10**, 4477–4509. (2017) https://doi.org/10.5194/gmd-10-4477-2017
4. Holdaway, D. and Kent, J.: Assessing the Tangent Linear Behaviour of Common Tracer Transport Schemes and Their Use in a Linearised Atmospheric General Circulation Model. Tellus A., **67**, 27895. (2015) http://dx.doi.org/10.3402/tellusa.v67.27895
5. Zarzycki, C. M., Jablonowski, C., Kent, J., Lauritzen, P. H., Nair, R., Reed, K. A., Ullrich, P. A., Hall, D. M., Taylor, M. A., Dazlich, D., Heikes, R., Konor, C., Randall, D., Chen, X., Harris, L., Giorgetta, M., Reinert, D., Kühnlein, C., Walko, R., Lee, V., Qaddouri, A., Tanguay, M., Miura, H., Ohno, T., Yoshida, R., Park, S.-H., Klemp, J. B., and Skamarock, W. C.: DCMIP2016: The Splitting Supercell Test Case. Geosci. Model Dev., **12**, 879–892. (2019) https://doi.org/10.5194/gmd-12-879-2019

Environment Modelling

Visualising Flood Frequency, Flood Volume and Mitigation of Extreme Events

O. Bokhove, M. A. Kelmanson, G. Piton, and J.-M. Tacnet

Abstract Two novel visualisation tools on extreme flooding are considered and their appeal discussed. First, an overview of the *Wetropolis Flood Demonstrator* is given. This is a portable set-up that showcases what a return period of an extreme flood event is: it has been received well by the general public. Second, a novel graphical cost-effectiveness diagnostic aimed at the public and decision-makers is explained: it has already led to increasing stakeholder participation in flood-mitigation planning.

1 Introduction

It is often difficult for the general public to understand the frequency and irregularity of extreme flooding events. In particular, what does it mean when a flooding event is denoted as a 1:100-year event or, equivalently, as an event with a 1% Annual Exceedance Probability (AEP)? Similarly, it is often difficult for them to grasp how much water is involved in the flooding of a city, as well as the efficacy and variety of flood-mitigation measures proposed to deal with this amount of water. For example, how can one compare the effects of steel-sheet-piling flood-defence walls to those of enhancement of upstream flood-plain storage? Vice versa, it is challenging for flood practitioners to explain to the public such uncertainties, as well as complementary grey and green engineering in flood-defence works, and their designs.

In attempting to bridge this gap in understanding and to minimise misunderstandings, we have devised new visualisation tools; one inspired by the other and both triggered by conversations with professional flood analysts. Neither of these tools

O. Bokhove (✉) · M. A. Kelmanson
Leeds Institute for Fluid Dynamics and School of Mathematics, Leeds, LS2 9JT, UK
e-mail: o.bokhove@leeds.ac.uk

G. Piton · J.-M. Tacnet
Univ. Grenoble Alpes, INRAE, CNRS, IRD, Grenoble INP, IGE, 38000 Grenoble, France
e-mail: guillaume.piton@inrae.fr

J.-M. Tacnet
e-mail: jean-marc.tacnet@inrae.fr

P. J. Aston (ed.), *More UK Success Stories in Industrial Mathematics*,
Mathematics in Industry 42, https://doi.org/10.1007/978-3-031-48683-8_5

replaces engineering calculations; rather they assist in both establishing scenario-based designs and communicating more finalised designs of flood-mitigation plans. The first tool is the *Wetropolis Flood Demonstrator*: a physical and portable miniature river set-up with a city intermittently flooded by simulated weather generated in a probabilistic manner, thus showing the meaning of return periods or AEPs of extreme rainfall and flooding events. The second tool is a graphical cost-effectiveness diagnostic of flood-mitigation measures that facilitates scenario-thinking for decision-makers and public planners. Research published in reports and articles [1, 2] evidences how these tools arose, and their societal impacts are summarised in [3]. Following overviews of the two tools, we focus on recently accrued highlights.

2 Wetropolis: From Flood to Climate and Drought Demonstrator

Visualisation of extreme flood-event return periods. The goal of Wetropolis is the visualisation of a return period or an AEP for an extreme flooding event. Its creation, propelled by requests from the Environment agency and JBA Trust, was to allow the public to see that a return period of 100 years, say, can occur at random intervals. Wetropolis reaches that goal in a portable model in which probabilistically generated weather leads to spatio-temporal rainfall onto an upland moor and/or reservoir intermittently entering the river on top of the constant upstream river inflow.

"Wetropolis days" (WDs) occur in 10 s of real time, thus rendering possible the visualisation of rare (e.g. AEP $= 1\%$) events without the audience having to wait over timescales of years. Rainfall durations are toggled by a ball first falling through a skewed Galton board that sets a duration of (1, 2, 4, 9)s with respective probabilities $(p_1, p_2, p_3, p_4) = (3, 7, 5, 1)/16$. Rainfall locations occur via a second board assigning rain in: (reservoir, moor and reservoir, moor, nowhere) with probabilities $(q_1, q_2, q_3, q_4) = (3, 7, 5, 1)/16$. The product of these discrete probabilities determines a probability matrix $P_{ij} = p_i q_j$ with $i, j = 1, 2, 3, 4$. The rare event concerns rainfall with the longest duration of 9 s, which has probability of $p_4 = 1/16$; the moor-and-reservoir location, with $q_2 = 7/16$, yields $P_{42} = 7/256 \approx 2.7\%$. Note that, for a dry day, rain-duration outcomes of the first Galton board are not used, such that no rain occurs with probability $\sum_{i=1}^{4} P_{i,4} = q_4 \sum_{i=1}^{4} p_i = q_4 = 1/16$. One second of rainfall in either reservoir or moor is driven by aquarium pumps each discharging a volume r_0 of water per second. Scaled by r_0, 1 s of WD rainfall has a chance of $p_1(q_1 + q_3)$, probabilities of rainfall flowrates $(0, 1, 2, 4, 8, 9, 18)r_0$ being:

$$\begin{aligned} &0 : q_4, \quad 1 : p_1(q_1 + q_3), \quad 2 : p_2(q_1 + q_3) + p_1 q_2, \\ &4 : p_3(q_1 + q_3) + p_2 q_2, \quad 8 : p_3 q_2, \quad 9 : p_4(q_1 + q_3), \quad 18 : p_4 q_2. \end{aligned} \tag{1}$$

The river channel is circa 4 m long, and there is upstream inflow of water of $10r_0$ with the moor halfway downstream of the river, a city at the downstream end of

Fig. 1 Photograph of the Wetropolis set-up (with Galton boards in the right-hand background) at the 2022 MPE exhibition in London. The reservoir is seen in the foreground as well as the square representing the city on the right. The porous moor is seen as the glass tank in the middle on the left: rainfall (from the pipe above the tank) and the water table can be discerned therein

the river and the reservoir between moor and city (Fig. 1). Based on a mathematical and numerical design model of a system of coupled partial and ordinary differential equations, the length of the day and the unit pump strength r_0 were determined such that major flooding occurs when $18r_0$ enters the set-up with an event probability of $p_e = P_{2,4} = 7/256$. Galton-board outcomes are coupled to matching rainfall using Arduino technology. Detailed modelling and design specifications are found in [2].

The return period T_r of a one-day flooding event uses the Wetropolis Event Probability $p_e = 2.7\%$ and $T_d = 10\,\text{s}$. When a one-day extreme event has happened on day zero, the chance p_n that it reoccurs on day $n > 0$ is given by a geometric distribution $p_n = (1 - p_e)^{n-1} p_e$. The mean (the expectation of $t_n = nT_d$) and standard deviation σ_r follow via manipulation of the geometric series and are respectively

$$T_r = \mathbb{E}(t_n) = \frac{T_d}{p_e} = 6{:}06\,\text{min}; \quad \sigma_r = \mathbb{E}\left((t_n - \mathbb{E}(t_n))^2\right) = \sqrt{(1 - p_e)} T_r = 6{:}00\,\text{min}.$$

Wetropolis was exhibited to over 1200 participants at the Mathematics of Planet Earth (MPE) exhibitions of 2020 and 2022, and a 7 m14 s educational video of it has been posted at https://youtu.be/rNgEqWdafKk, thereby facilitating more reach [6].

Probability of super- and mega floods, climate and drought. Superfloods occur in Wetropolis when extreme rainfall of 9 s/WD in moor and reservoir happens on two consecutive days. Intuitively, this has a return period of $T_r^{(2)} = T_d/p_e^2$. However, return periods and standard deviations for such rainfall on k consecutive days have been calculated exactly using a geometric distribution of order k, yielding:

$$T_r^{(k)}/T_d = \frac{(1-p_e^k)}{(1-p_e)p_e^k}, \quad \sigma_r^{(k)}/T_d = \frac{\sqrt{1-(2k+1)(1-p_e)p_e^k - p_e^{2k+1}}}{(1-p_e)p_e^k}. \tag{2}$$

For small p_e, at leading order $T_r^{(k)} \approx T_d/p_e^k$, which is exact for $k=1$, and $\sigma_r^{(k)} \approx T_r$.

For the extreme event with $p_e = 7/256$, the return periods for $k = 2, 3$ thus become $T_r^{(2)} = 3.8\,\text{h}$ and $T_r^{(3)} = 140\,\text{h}$. The two-day superflood makes quite an impression on the audience and has been seen and filmed a few times. The three-day megaflood has not been seen hitherto. The one-day flood event is weaker and sometimes absent if the set-up has not been balanced properly. By changing the event probability to $p_e = 49/256$, so $(7/16)^2$ from both Galton boards, the stronger super- and megafloods with return periods $T_r = 52$ s, $T_r^{(2)} = 5{:}25$ min and $T_r^{(3)} = 29.11$ min become feasible with the disadvantage that 9 s rainfall becomes most frequent. Alternatively, one could consider a WD of duration 20 s such that rainfall can vary per day.

Climate change has been included and showcased in Wetropolis by adding an extra lake upstream of the moor, with rainfall in this lake and moor triggered by the moor's Galton-board outcome, with a weaker pump for the lake than for the moor. This leads to an increase of $\sim$20% in average rainfall under climate change, and the one-day flooding of the city thus becomes more severe and noticeable. Droughts can be enhanced by allowing a subset of the no-rain days to last longer, say four days, with modified probabilities. A drought can be visualised in the set-up by a drinkwater pipe to the city drying up because, in the absence of rainfall, the moor's water table will sink below an inlet point in the moor. Taken collectively, the above considerations clearly begin to meet requests to link Wetropolis' rainfall to actual weather data, and to illustrate climate change and drought.

3 Graphical Cost-Effectiveness Diagnostic Tool on Flooding

Though the aftermath of the 2015 Boxing Day flood in Leeds, UK—and the resulting evacuation of a business in the River Aire's flooded plain—led to the inception of Wetropolis, it was inspection of the subsequent public 2017 Flood Alleviation Scheme II (FASII) plans of Leeds City Council (LCC) that triggered a novel flood-mitigation cost-effectiveness diagnostic. This new tool allowed succinct visualisation and summary of the FASII plan and concomitantly identified some inconsistencies and missing information that were reported to LCC. Further public archiving of the general tool allowed its rapid and wider use within the EU, usage later corroborated in the REF2021 impact case study [3].

The tool is summarised schematically in Fig. 2a, b and c. In sub-figure (a) is a three-panel graph in whose third quadrant directly measured water levels $h(t)$ (axis to left) are displayed as function of time t (axis downwards). Discharge Q (vertical axis shared with quadrant one) is obtained via a (blue) rating curve, alternatively $h(Q)$ or $Q(h)$, in quadrant two. In quadrant one, the discharge $Q(t)$ follows as a

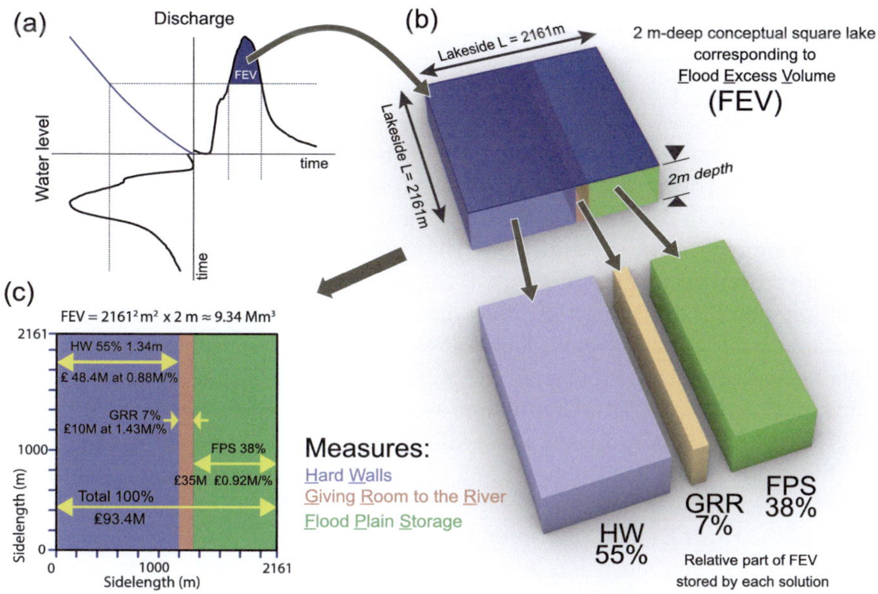

Fig. 2 Schematic of graphical cost-effectiveness diagnostic tool. Abbreviations in figure or text

function of time (new horizontal axis to right). Given a water-level threshold (right vertical line in quadrants two and three), above which intermediate-to-major flooding occurs around the measurement site, a discharge threshold (lower horizontal line in quadrants two and one) follows. It demarcates the blue area, labelled FEV to represent flood-excess volume; FEV is the fraction of the flood volume that causes the flood damage. Using this approach, the FEV for Leeds' 2015 Boxing Day flood emerged as $\mathrm{FEV}_l = 9.34\,\mathrm{Mm}^3$: in order to contextualise such a volume of water, FEV is herein expressed as a square lake with human-size (shallow) depth $D = 2\,\mathrm{m}$ and side-length $L = \sqrt{\mathrm{FEV}/D}$. Such a lake is shown in sub-figure (b) and, since $L \gg D$ for intermediate-to-large floods (here $L = 2161\,\mathrm{m}$), the vertical scale is exaggerated in the figure. Flood mitigation of the entire FEV amounts to partitioning the square lake into rectangular or quadrilateral shapes for each mitigation measure. For illustration, the three measures chosen are higher flood-defense walls (HW), giving-room-to-the-river (GRR – effectively lowering and widening the river borders so that they convey more water before flooding the site) and flood-plain storage (FPS – enhancing the storage volume of the flood plain by operating a dynamic weir to further stow up flood waters). In sub-figure (c) the square lake is plotted from above with, for each mitigation measure, its costs, costs per percentage (here under linear scaling) as well as total costs. Refinements can be made by including uncertainties and errors in rating curves, water-level measurements and mitigation measures; it is accordingly accepted that FEV will generally incur potentially large errors. Such uncertainties can readily be shown in the square-lake graphs by using quadrilateral shapes with worst-case values at the bottom and best-case ones at the top; similarly spatially varying rainfall scenarios and mitigation measures can be included (e.g. [3, 4] and references therein). The diagnostic tool is based on a calculated or measured flood hydrograph at one location and therefore appears to be spatially zero-dimensional, but in practice it represents a single stretch of river. Accordingly, the diagnostic can be extended to involve multiple hydrographs from a series of linked river stretches.

With evidence to corroborate its usage [3], the diagnostic tool can and has been applied in three types of investigation: (i) as a consistency check of flood-mitigation plans with proven (as yet uncorroborated) ability to uncover inconsistencies and missing information [3, 4]; (ii) as an a priori investigation to define mitigation scenarios before any detailed engineering calculations are done or when these are too costly; as such allowing communication about the physical capacity of measures, stakeholder participation and dismissal of mitigation measures with undersized FEV capacity (as performed by an NGO for the River Glinščica in the Ljubljana, Slovenia [3, 4]); and, (iii) as an a posteriori executive summary of detailed engineering calculations of flood-mitigation plans, as done for a French municipality after devastating floods in 2015 of the River Brague. Consultants picked up on the River Brague case report [5] to perform a comprehensive study for detailing an appropriate GRR-mitigation measure that was first raised by our FEV-cost-effectiveness diagnostic.

Finally, we have introduced the concept of *available flood-storage volume* (see references in [3]) which aids in assessing the efficacy of FPS. It has both highlighted unclear aspects of FPS within Leeds' FASII plan and demonstrated that the efficacy of FPS within (much-publicised) beaver ponds is small, as now quantified. Consider an idealised river with a one-sided flat flood plain of width $W = 100\,\text{m}$ and length $L_f = 1\,\text{km}$ with an adjacent rectangular river channel of depth d. For a 1:100yr flood the flood-plain depth is $d_1 = 1\,\text{m}$ and for a 1:300yr one the depth $d_4 = 4\,\text{m}$, say. A new weir can dynamically raise the flood-plain level to $d_w = 4\,\text{m}$. The available flood-plain storage volume for protection against the 1:100yr flood is then $V_{afps} \le (d_w - d_1)L_f W = 3 \times 10^5 m^3$ and, similarly, $V_{afps} \le (d_w - d_4)L_f W = 0\,\text{m}^3$ against the 1:300yr flood. Estimation of V_{afps} for a series of dams of a beaver colony in Devon is illustrative. During a rainfall event, the water levels of the beaver ponds can be seen to fluctuate in a limited manner with $V_{afps} \in [200, 1100]\text{m}^3$ for the beaver colony (the $200\,\text{m}^3$ estimate arises in [5]). For perfect dams, upscaling to reach 10% protection for Leeds' $\text{FEV}_l = 9.34\,\text{Mm}^3$ would then require $\text{FEV}_l / V_{afps} \in [849, 4245]$ beaver colonies (with the other 90% of FEV covered otherwise), which is unrealistic [1, 5]. Hence, overstated promotion of beaver colonies as a flood-mitigation measure draws attention away from their actual and dominant benefit: wildlife enhancement.

References

1. Bokhove, O., Kelmanson, M, Kent, T.: Using flood-excess volume in flood mitigation to show that upscaling beaver dams for protection against extreme floods proves unrealistic. https://eartharxiv.org/w9evx/ (2018)
2. Bokhove, O., Hicks, T., Zweers, W. Kent, T.: Wetropolis extreme rainfall and flood demonstrator: from mathematical design to outreach and research. *Hydro. Earth Syst. Sc.* **24(5)**, 2483–2503. https://doi.org/10.5194/hess-24-2483-2020 (2020)
3. Bokhove, O., Kelmanson, M., Hicks, T., Kent, T.: Flood mitigation: from outreach demonstrator to a graphical cost-effectiveness diagnostic for policy makers. REF2021 Impact Case Study, https://results2021.ref.ac.uk/impact/submissions/1eedb5bd-8f92-4737-a6f0-1e61c997e4f0/impact
4. Bokhove, O.: On communicating cost-effectiveness of flood-mitigation schemes. ESREL conference Angers, France, https://www.rpsonline.com.sg/proceedings/9789811820168/pdf/134.pdf (2021) (2022)
5. Piton, G., Dupire, S., Arnaud, P., Mas, A., Marchal, R., Moncoulon, D., Curt, T., Wang, Z-X., Onfroy T., Tacnet, J-.M.: From hazard to risk: models for the DEMOs NAIAD (Nature insurance value: assessment and demonstration). Part 6 France Brague catchment. 215 - 344 http://naiad2020.eu/wp-content/uploads/2019/02/D6.2_REV_FINAL.pdf (2018)
6. Wetropolis video Leeds Institute for Fluid Dynamics: https://www.youtube.com/watch?v=yUjYfg2SfY0 (2024)

Improving National Water Quality Monitoring and Network Design Through Statistical Methods

Claire Miller and Marian Scott

Abstract Environmental monitoring of fresh surface waters is essential to protect human and animal health and ensure compliance with regulation. There are ongoing challenges around where and when to monitor water quality in order to meet these requirements, take advantage of new sensor technology and balance complex budgetary constraints. In this work, statistical modelling tools based on generalised additive models, functional data analysis and multivariate methods enabled interrogation of historical monitoring data and the ability to assess the capability of current and future monitoring network designs to provide evidence on water quality trends. Evidence from such statistical and data analytic approaches enabled assessment of compliance with water quality monitoring regulation and informed the design planning for monitoring networks, with a view to improving water quality monitoring within practical and budget constraints.

1 Introduction

Fresh surface waters are regularly monitored for compliance and research purposes in order to investigate the water quality across space and time, explore what drives these patterns and to ensure compliance with water quality directives, in order to protect human and animal health. The European Union Water Framework Directive 2000/60/EC established an integrated framework for the protection of surface waters and each member state is required to demonstrate compliance through statutory reporting of water quality monitoring. The UK operated under these directives until recently, and beyond Brexit regulation and compliance to maintain/improve water quality continues through, for example, the UK government 25-year environment plan.

C. Miller (✉) · M. Scott
School of Mathematics and Statistics, University of Glasgow, Glasgow G12 8QQ, UK
e-mail: claire.miller@glasgow.ac.uk

M. Scott
e-mail: marian.scott@glasgow.ac.uk

P. J. Aston (ed.), *More UK Success Stories in Industrial Mathematics*,
Mathematics in Industry 42, https://doi.org/10.1007/978-3-031-48683-8_6

Traditionally, water quality samples are taken manually at, for example, weekly to monthly intervals from within freshwater lakes and rivers. Such samples are analysed in regulated laboratories to determine the level of, for example, nutrient concentrations. Statistical models can be used to: investigate patterns over space and time, explore relationships between nutrient concentration levels and variables used as proxies for water quality (such as chlorophyll$_a$), and estimate uncertainty around parameter estimates from such statistical models. There are on-going challenges around the future design of environmental monitoring networks as a result of resource/budgetary constraints. However, there is also the potential to improve such networks through the use of new sensor technology. Statistical principles and developments can help to inform the selection of future monitoring sites and/or the future location for placement of automatic sensors.

In our previous work, spatiotemporal statistical methods have been developed to investigate trends over space and time in water quality data for freshwater applications. Such approaches enable data-driven nonlinear patterns to be estimated to describe changes in responses of interest over space and time, for example to efficiently estimate temporal trends and seasonal patterns in water quality [1]. Similar methods were developed and applied collaboratively with the Environment Agency in [2] to efficiently investigate spatiotemporal trends in river water quality in the UK. Additionally, in order to plan for future water management and monitoring, it can be useful to investigate such nutrient spatiotemporal patterns in order to identify where similar patterns occur over time and spatially across a river monitoring network. Identification of dominant spatial and temporal patterns can indicate time points where water quality has remained stable over the network or identify groups of monitoring sites that exhibit similar temporal patterns [3].

This article provides a summary of the methods developed in our work to describe spatial and temporal patterns of nutrient concentration levels within river networks, and to investigate the presence of common patterns over space and time to help inform the design planning of future monitoring networks.

2 Spatiotemporal Additive Models

River water quality monitoring networks can include hundreds of in-situ monitoring sites placed across the network (Fig. 1) at which samples are taken and analysed over time for levels of nutrient concentrations e.g. nitrogen and phosphorus. Spatiotemporal additive models have been a useful way to identify the spatial and temporal patterns in nutrients, with a hierarchy of models being of interest, including:

$$y_{ij} = \beta_0 + f_1(\text{easting}_{ij}, \text{northing}_{ij}) + f_2(\text{year}_{ij}) + f_3(\text{week}_{ij}) + \epsilon_{ij}, \qquad (1)$$

where y_{ij} is a response of interest at time ($i = 1, \ldots n$) and site ($j = 1, \ldots, p$), f_k represent the k smooth functions determined typically using splines or kernel smoothing for $k = 1, \ldots, 3$, easting and northing represent the spatial locations for

Fig. 1 Flowing river water (left) with example flow-connected network (right) with flow direction indicated by arrow and monitoring sites indicated by black dots

each site, year typically represents a continuous variable over time, week represents a cyclic component to explore seasonal patterns, and $\epsilon_{ij} \sim N(0, \sigma^2)$. Additional higher order interaction terms enable the fitting of spatial and/or seasonal patterns that change over time.

Models such as Eq. (1) are often fitted using penalised regression splines and/or kernel smoothing and can be extended as appropriate to include more complex error structures in order to robustly estimate the uncertainty associated with fitted effects. In particular, in the scenario of river water, the physical connections between monitoring sites, due to flow, the tree-like structure of rivers and regular temporal frequency of monitoring means that there potentially exists a complex spatial and temporal correlation structure for measured responses.

The output from models such as Eq. (1) enables a description of the long-term trend in the response (e.g. nutrient levels), the estimated seasonal pattern (which can be allowed flexibly to change over time), and the estimation of the pattern over space (which can also be allowed to flexibly change with time). The research in [2] evaluated the changes in nutrient patterns (based on nitrogen/phosphorus weekly to monthly measurements) in river water over time (over a 20–40 year period up until 2010) and spatially for each of 59 large hydrological areas in England & Wales, to identify evidence of historic trends in the concentration of nutrients in rivers, and to describe variations between catchments.

3 Space-Time PCA and Clustering

If we can identify common spatiotemporal patterns, within e.g. nutrient levels of water from river networks, then potentially these common patterns can be used as one approach to inform the design of future monitoring networks. Our work in [3] presents an approach to investigate identification of common spatiotemporal patterns within river network nutrient data. This specifically uses the river network structure in order to improve estimation of common temporal patterns by incorporating information on the flow direction of the river and monitoring site connectedness (e.g. where water from site 1 flows to site 2 but site 2 does not flow to site 1, Fig. 1 (right)).

The nutrient levels within river water quality recorded on days that are close in time are highly likely to be related to one another through common seasonal, catchment and location effects. Nutrients are dispersed through diffuse or point source pollution and hence, concentration levels between river monitoring locations that are flow-connected are highly likely to be more similar than those that are unconnected further upstream on different tributaries, and common catchment land types and influences or drainage areas can also result in monitoring locations with similar nutrient levels. Statistically, these features are referred to as temporal and spatial correlation respectively. Such correlation can mask data features i.e. it can be difficult to identify influential/dominant sites and/or time points in the presence of such correlation.

Principal Components Analysis (PCA) is typically used for dimensionality reduction in the presence of correlated variables to produce independent weighted linear combinations of the original variables which (in a smaller dimension) explain a large proportion of the variability in the data. In the presence of measurements over space and time for a single variable, PCA can be used to identify dominant spatial and/or temporal patterns. Additionally, PCA can be modified using weights to remove additional layers of spatial and/or temporal correlation (i.e. whitening) to unmask hidden structure. For river networks, weights reflecting the upstream monitoring site influence on downstream sites can be calculated using watershed area (a proxy for discharge), where the watershed area is the area of land draining directly to a stream segment. Appropriate row and column weights can be used to adjust PCA for structure, resulting in a new matrix $\tilde{\mathbf{X}}$:

$$\tilde{\mathbf{X}} = \mathbf{\Phi}\mathbf{X}\mathbf{\Omega},$$

where $\mathbf{X}$ is an $n \times p$ original covariance/correlation matrix used for PCA (with n data points for each of p variables), $\mathbf{\Phi}$ and $\mathbf{\Omega}$ are row and column weight matrices of dimension $n \times n$ and $p \times p$ respectively. PCA is then applied to $\tilde{\mathbf{X}}$ and the PCs are then related to $\tilde{\mathbf{X}}$ instead of $\mathbf{X}$, with the loadings and PCs for $\mathbf{X}$ produced through back transformation. This approach, referred to as flow-directed PCA, is fully described in [3].

In flow-directed PCA, the rows of $\mathbf{X}$ are the n time points for each of the p columns of spatial locations, $\mathbf{\Phi}$ can be used to incorporate temporal weights and $\mathbf{\Omega}$ spatial weights. The spatial weights ($\mathbf{\Omega}$) are determined by the area of land draining to a stream segment and monitoring site connectedness. Temporal weights ($\mathbf{\Phi}$) can be specified by using appropriate time series ARMA structures.

The output from flow-directed PCA indicates the most influential monitoring sites within the network (through the sites containing the largest PC loadings). Additionally, clustering these PC loadings (e.g. through a k-means approach) can identify monitoring sites where nutrient patterns are similar (and/or show the highest level of variation) over time. See [3] for more details, application of the approach and related software.

4 Impact

The statistical and data analytic approaches described above have been used by environmental regulators to inform the evidence base in support of UK policy regulating environmental water quality, and to inform the planning for the design of new monitoring networks, where there is a requirement to balance practical and budgetary constraints while improving the spatial representativeness of any network.

In order to design a monitoring network there are several key considerations including: the purpose of the monitoring network, the questions of interest to be investigated, and the spatial and temporal scale of the information that is required. All of this will directly impact the inferences that can be derived from the data recorded across the monitoring network. In recent years, design of monitoring networks has been a key consideration within environmental applications and several environmental regulators have been investigating the potential of the generalised random tessellation survey (GRTS) approach [4]. The approach is commonly applied in an environmental context, due to its ability to produce a spatially representative sample of the wider population. The approach can be used to develop a spatially representative design of monitoring sites across an area with a repeated measurement design over time to make inferences about spatial trends, status and temporal trends from the one design. However, the power to detect trends and describe status will depend on the balance of sites across space and repetition across time, in addition to the signal to noise ratio of the measured determinand of interest. Related approaches to those described in this article have been used to initially investigate choices and implications for such designs.

Our current research includes developing a framework and principles for environmental sensor network design and developing methodology for data fusion [5] (the technique of combining multiple data streams of differing spatial and temporal support using statistical/data analytic/deep learning approaches). For example, combining all relevant data sources for the same freshwater determinand from (say) in-situ monitoring networks, automatic sensors and satellite data will provide the most complete picture possible to make inferences around water quality estimation and prediction.

Acknowledgements The authors would like to thank the funders and colleagues involved with related research to this work. The Environment Agency and the Scottish Environment Protection Agency for collaborations on related work. The Environment Agency for partial funding of related work. Kelly Gallacher who was funded by an Engineering and Physical Sciences Research Council DTA and the EPSRC SECURE network (EP/M008347/1), Craig Wilkie who was funded by an EPSRC Impact Acceleration Award (EP/R511705/1) and Marnie Low who was funded by the Environment Agency and EPSRC Impact Acceleration Award (EP/R511705/1).

References

1. Ferguson, C.A., Carvalho, L., Scott, E.M., Bowman, A.W. and Kirika, A.: Assessing ecological responses to environmental change using statistical models. Journal of Applied Ecology. **45**, 1, 193–203 (2008)
2. Miller, C., Magdalina, A., Willows, R., Bowman, A., Scott, E., Lee, D., Burgess, C., Pope, L., Pannullo, F., and Haggarty, R.: Spatiotemporal statistical modelling of long-term change in river nutrient concentrations in England & Wales. Science of the Total Environment. **466-7**, 914–923 (2014)
3. Gallacher, K., Miller, C., Scott, E.M., Willows, R., Pope, L., Douglass, J.: Flow-directed PCA for monitoring networks. Environmetrics. **28**, e2434 (2017)
4. Olsen, A.R., Kincaid, T.M., Payton, Q.: Spatially balanced survey designs for natural resources. Design and Analysis of Long-term Ecological Monitoring Studies. CUP (2012)
5. Wilkie, C.J., Miller, C.A., Scott, E.M., O'Donnell, R.A., Hunter, P.D., Spyrakos, E., and Tyler, A.N.: Nonparametric statistical downscaling for the fusion of data of different spatiotemporal support. Environmetrics. **30**, 3, e2549 (2019)

Modelling the Rise of Volcanic Plumes to Estimate Mass Eruption Rates

Andrew J. Hogg and Mark J. Woodhouse

Abstract Explosive volcanic eruptions can generate plumes that carry ash high into the atmosphere posing hazards to aviation while airborne and to lives and infrastructure as the ash settles back to ground level. It is crucial to quantify the rate at which ash is produced if its subsequent dispersion is to be modelled accurately. Direct measurement of this mass eruption rate is impossible; however, observations of plume dynamics provide a means of determining this quantity.

1 Introduction

Explosive volcanic eruptions inject ash into the atmosphere and its subsequent dispersion and deposition causes major hazards. The very largest eruptions have transported such large volumes of ash into the stratosphere that global temperatures and the climate are impacted. However even smaller events pose hazards to health and infrastructure as the erupted ash settles back to ground level. Airborne ash also poses a danger to aircraft flights because modern jet engines are susceptible to damage, even from low concentrations. This hazard gained particular notoriety during the 2010 eruption of Eyjafjallajökull, Iceland, during which European airspace was closed for several weeks with the cancellation of over 100,000 flights and the direct economic loss across all sectors estimated as US$5bn.

Prior to the 2010 Eyjafjallajökull eruption, the International Civil Aviation Organization adopted a precautionary policy of ash avoidance, with no concentration of ash in the atmosphere considered safe for aircraft. However, the disruption to aviation during the first week of explosive activity at Eyjafjallajökull (14–18 April 2010) led

A. J. Hogg (✉)
School of Mathematics, University of Bristol, The Fry Building, Woodland Rd, Bristol BS8 1UG, UK
e-mail: a.j.hogg@bristol.ac.uk

M. J. Woodhouse
School of Earth Sciences, University of Bristol, Wills Memorial Building, Queen's Road, Bristol BS8 1RJ, UK
e-mail: mark.woodhouse@bristol.ac.uk

P. J. Aston (ed.), *More UK Success Stories in Industrial Mathematics*,
Mathematics in Industry 42, https://doi.org/10.1007/978-3-031-48683-8_7

to a relaxation of this policy in Europe, with the UK Civil Aviation Authority and Eurocontrol introducing ash concentration thresholds for commercial air traffic. Ash concentrations below 2mg m^{-3} are considered safe for flights, while flight operations at higher concentrations require a Safety Case to be accepted by national regulators. Typically, Safety Cases have been accepted for ash concentrations up to 4mg m^{-3}. The introduction of thresholds of ash concentration places increased demands on accurate modelling of atmospheric ash dispersion during volcanic crises.

The crucial component for forecasting ash dispersion is the estimate of the strength of the volcanic source. In other words, what is the rate of production of ash from the volcano? Direct measurements of this mass eruption rate (MER) are extremely difficult due in part to the explosive character of the volcanic dynamics at the vent, and so the MER is typically determined through interpretation of indirect measurements. In this paper we report on mathematical modelling that is used to interpret observations of volcanic plumes and to determine the MER, and we indicate how this model has been used in operational settings.

Volcanic plumes rise through the atmosphere because the erupted ash is hot and rapidly heats the surrounding gases to generate a buoyant release. The resultant plume ascends and mixes with the denser atmospheric air, increasing its bulk density. The atmosphere through which it rises is usually stably stratified (its density decreases with height), and so eventually the plume reaches altitudes at which it is no longer less dense than the surroundings and begins to spread horizontally. The height of rise, therefore, is related to the strength of the source (i.e. the MER), as well as other atmospheric properties such as the stratification and the wind field.

Data from historic eruptions for which there are observations of the height of rise through the atmosphere, H, and estimates of the MER, Q, indicate that there is approximately a $\frac{1}{4}$-power relationship between the two. For example, [1] propose the empirical fit, $H = 0.220Q^{0.259}$, where H is measured in kilometers and Q in kilograms per second. This relation has been used operationally to infer the MER from observations of the height of rise of the plume. Though simple, this approach is problematic because the historical data is bias towards the very largest eruptions for which it is easiest to estimate the erupted volume, and because the formula neglects the effects of atmospheric conditions, which will be shown to be important; critically for the relatively small-scale eruption of Eyjafjallajökull in 2010, the empirical relation takes no account of the effects on the plume of the atmospheric wind.

2 Mathematical Model of Volcanic Plumes

Integral models of turbulent plumes were first developed by [2]. The approach entails averaging over the fluctuating turbulent characteristics to produce models that capture the steady vertical variation of the plume properties, such as its velocity, density and cross-sectional area. The key effect of turbulence is to induce mixing between the rising buoyant fluid and the surroundings. This mixing is embodied into an entrainment velocity that captures that rate at which atmospheric fluid is engulfed into the plume [2]. Integral models have the distinct advantage over more complicated simulations of the fluid flow that they are very simple to compute (a vital feature for operational hazard assessment), and yet they have been shown to capture features such as the height of rise accurately. Our model extends the theory of turbulent plumes [2] to volcanic settings, in which there is non-negligible wind. The wind bends over the ascending plume (see Fig. 1) and modelling equations are developed as functions of the arc-length, s, such that the vertical position, $z(s)$ and the horizontal position, $x(s)$ (where the x-axis is aligned with the atmospheric wind), are given by

$$\frac{\mathrm{d}x}{\mathrm{d}s} = \cos\theta \qquad \text{and} \qquad \frac{\mathrm{d}z}{\mathrm{d}s} = \sin\theta, \tag{1}$$

where $\theta(s)$ is the angle between the axis of the plume and the horizontal.

The model mathematically embodies four physical principles to derive equations for temperature, $T(s)$ (and hence density $\rho(s)$), axial velocity, $U(s)$, inclination, $\theta(s)$ and cross-sectional area, assumed circular with radius, $R(s)$, given by

$$\frac{\mathrm{d}}{\mathrm{d}s}\left(\rho U R^2\right) = 2\rho_a U_e R, \tag{2}$$

$$\frac{\mathrm{d}}{\mathrm{d}s}\left(\rho U^2 R^2 \sin\theta\right) = (\rho_a - \rho)\, g R^2, \tag{3}$$

$$\frac{\mathrm{d}}{\mathrm{d}s}\left(\rho U^2 R^2 \cos\theta\right) = 2\rho_a U_e R V, \tag{4}$$

$$\frac{\mathrm{d}}{\mathrm{d}s}\left(\rho U R^2 \left(C_p T + \frac{U^2}{2} + gz\right)\right) = 2\rho_a R U_e \left(C_{pa} T_a + \frac{U_e^2}{2} + gz\right) \tag{5}$$

These equations express mass conservation, momentum balance in the vertical and horizontal and energy conservation, respectively. Quantities with subscript a refer to atmospheric variables, V is the wind speed, g gravitational acceleration and C_p the specific heat capacity. Volcanic ash is assumed to be conserved during plume ascent

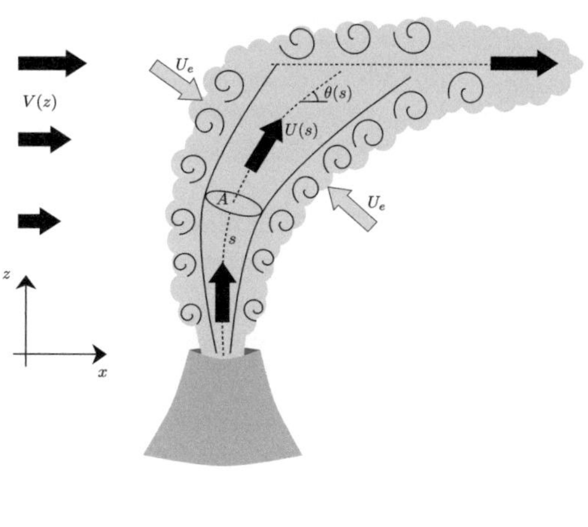

Fig. 1 Schematic of a volcanic plume in a crosswind and the variables used in the mathematical model of the motion

because its settling is much slower than the plume velocity. Gases are modelled using ideal gas laws, with properties that vary with the solids mass fraction (see [3]). The final important closure relates the entrainment velocity, U_e, to the axial speed and the wind speed. This has been determined empirically as

$$U_e = k_s|U - V\cos\theta| + k_w|V\sin\theta|, \tag{6}$$

where k_s and k_w are known constants [3]. Importantly this formula connects the enhanced rate of mixing to the wind speed.

The governing Eqs. (1)–(5) may be integrated numerically from prescribed conditions at the volcanic vent to determine the maximum height of rise. It is possible to have source conditions that do not generate a buoyant plume, because there is insufficient heat transfer from the erupted material to the entrained air to reduce the density of the mixture so that it becomes buoyant. In this case the volcano issues an initially vertically-aligned jet that reaches its maximum height quite close to the source. However if the plume becomes buoyant then it ascends into the high atmosphere, driven by density differences rather than its initial momentum.

The model (1)–(5) provides a quantitative link between the source conditions (and hence the MER) and the maximum height of rise, and depends upon the atmospheric properties such as the vertical variation of temperature T_a, and the wind speed, V. We discuss the implementation of the model below, but first highlight some results.

2.1 Results

The most important result is that wind increases mixing and that this leads to lower heights of rise for otherwise identical explosive eruptions. This is most easily demonstrated by applying the model to analyse the height of rise through a US-standard atmosphere in which the temperature decreases linearly with height within the troposphere, while the wind increases with height reaching a maximum V_1 at the tropopause (see [3]). We integrate (1)–(5) for various values of the wind speed, V_1 and varying source conditions (see Fig. 2). The model results are consistent with the data, although there are inevitably some differences because the atmospheric conditions vary for each eruption. Importantly, though, the predicted height of rise decreases as the wind speed, V_1, increases. This has crucial consequences.

We used our model to interpret observations of the 2010 eruption of Eyjafjallajökull. During the first explosive phase of the eruption, the plume was observed to rise to approximately 8km on 14 April 2010, before falling back to 5km on 15 April and then rising again to over 8km on 16–17 April. One interpretation of this variation is that the volcanic eruption is varying its intensity (and MER), and this is the only possible quantitative explanation if one employs the empirical relation between height of rise of MER. However the wind speed (measured at an elevation of 7km) also varied during this period from relatively low speed (25ms^{-1}) to a much higher speed (50ms^{-1}) before falling back to below 20ms^{-1}. Thus an alternative

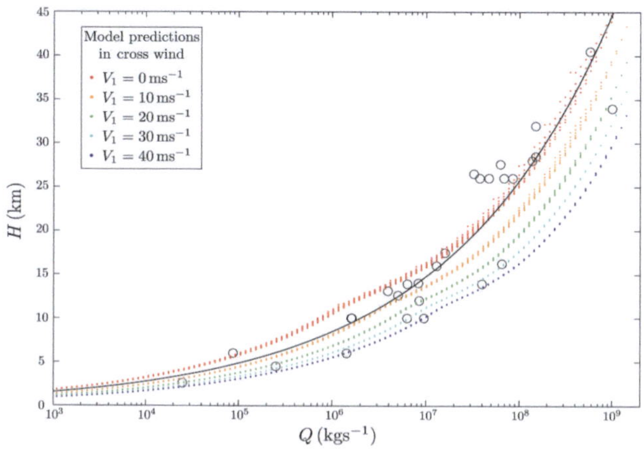

Fig. 2 The height of the plume rise, H, as a function of the mass eruption rate, Q, showing field data (o), the empirical curve of [1] (solid line), and the results from simulations with varying atmospheric wind speeds at the tropopause (coloured symbols)

explanation for the fluctuating plume rise is that the volcano is erupting at a constant rate throughout, but that the variation in wind speed is leading to the variation in plume height; this explanation is supported by our model analysis. Furthermore our calculations suggest that the MER is 6×10^6kg s^{-1}, far in excess of the estimate of 2×10^5kg s^{-1} that would be calculated using the Sparks-empirical relation when the plume had risen to it lowest height [3].

2.2 *Implementation*

The mathematical model is implemented within free-to-use software, PlumeRise, available at www.plumerise.bristol.ac.uk. This software integrates the model numerically using measured atmospheric conditions, as well as chosen values of the other parameters to determine the maximum height of rise. In addition it may be used as an inverse model to determine the MER given an observed height of rise. All computations are performed using the online server; the user has only to specify the source conditions, atmospheric profiles, and parameter values.

3 Impact

PlumeRise was launched in 2014 and is widely used by the community that has responsibilities for assessing the hazards to aviation. PlumeRise has over 500 registered users and over 100,000 separate calculations have been made in different eruptive scenarios from around the globe. These include over 170 practitioners and researchers in international and non-governmental agencies, including representatives from 8 of the 9 Volcanic Ash Advisory Centres (which have global responsibility for assessing the safety of aircraft flights).

The mathematical model of the rise of volcanic plumes and it implementation as PlumeRise have led to direct improvements in the operational assessment of environmental risk from volcanic ash during eruptive events and in contingency planning, which for the aviation industry, has the associated economic impact of avoiding unnecessary airspace closures while maintaining safe flying. Additionally, the research has impacted the use and development of quantitative plume models within several international agencies, including the quantification of uncertainty.

An example of its usage was during the long-lived Holuhran-Bardarbunga eruption, Iceland (2014–15). Although this event predominantly produced ground-based lava flows, the Icelandic Civil Protection and Emergency committee daily examined the possible consequences of an ash-producing eruption. The Icelandic Meteorological Office (IMO) held responsibility for monitoring and forecasting activities, both for Iceland, but also internationally. The IMO used PlumeRise operationally to compute source conditions for their dispersion models, and during the long-lived eruption, twice daily predictions of the potential heights of rise of the volcanic plume for a range of eruption rates were computed using the latest meteorological data; predictions that were shared with the committee for Icelandic Civil Protection.

Airborne ash additionally has consequences beyond aviation, affecting population health and livelihoods and the research has impacted international procedures for hazard assessment. In Mexico, for example, over 26 million people live within 100km of Popocatépeti Volcano and are affected by ash fallout from its frequent eruptions during the past three decades. Ash dispersion is modelled quantitatively by the Centro de Ciencias de la Atmósfera and the national civil protection agency of Mexico (Centro Nacional de Prevención de Desastres) to forecast the hazard to communities and the MER determined by [3] is a vital input to these computations.

Finally we remark that the 2010 eruption of Eyjafjallajökull challenged the international community to develop better forecasting tools and PlumeRise was part of the response. Best practice for plume modelling has been recently reviewed by International Association of Volcanology and Chemistry of Earth's Interior and its recommendations underpin hazard assessment worldwide. Continued development of models will be a critical component of strengthening mitigation measures.

Acknowledgements The authors acknowledge the financial support of NERC (NE/I015612/1), the EU (FP7-Environment, 308377) and an EPSRC Impact Acceleration Award.

References

1. Sparks, R.S.J., Bursik, M.I., Carey, S.N., Gilbert, J.S., Glaze, L.S., Sigurdsson, H. and Woods, A.W. (1997), Volcanic Plumes, John Wiley & Sons, Chichester, U.K
2. Morton, B.R.,Taylor, G.I. and Turner, J.S. Turbulent gravitational convection from maintained and instantaneous source. Proc. R. Soc. (1956), **234**, 1–23, https://doi.org/10.1098/rspa.1956.0011.
3. Woodhouse, M.J., Hogg, A.J., Phillips, J.C. and Sparks, R.S.J. Interactions between volcanic plumes and wind during the 2010 Eyjafjallajökull eruption, Iceland. (2013) J. Geophys. Res. **118**, 92–109, https://doi.org/10.1029/2012JB009592

Engineering

Danger on the Seas: Assessing the Risk to Ships From Wave Slamming Events

Alexander Korobkin, Tatyana Khabakhpasheva, and Sime Malenica

Abstract Large ships and offshore structures face a serious risk of damage from severe wave events in rough waters. Slamming events occur when the hull of a ship slams the water surface, or a wave slams a ship, potentially causing catastrophic damage. Simulations of slamming are complicated because large structures are flexible, and the slamming loads depend on the structural deformations. Computations of hydroelastic slamming require a structural code to evaluate stresses in the structure and assess the risk of the structural damage, a hydrodynamic code to determine the slamming loads, and a coupling interface between these codes. The weakest element in this scheme of the risk assessment, which is used by certification companies on a daily basis, is an interface code. It is shown below that in slamming simulations, where the added-mass effects play the leading role, construction of such an interface is not straightforward. To be confident in the results of computer simulations of hydroelastic slamming, the computed results should be very carefully validated. This validation can be done, in particular, by using analytical, fully coupled models of slamming. As a result, the risk assessment becomes more reliable leading to safer ship operations at sea.

1 Introduction

The safe operation of ships is of high priority in order to protect the ship, the personnel, the cargo and the wider environment. Ships and offshore structures, such as rigs and Floating Production Storage and Offloading Systems (FPSOs), risk serious damage from violent events in high seas. During slamming the hull of a ship collides rapidly with the water surface generating dangerously large local hydrodynamic loads. These loads may also produce violent vibrations of the ship, a phenomenon

A. Korobkin (✉) · T. Khabakhpasheva
University of East Anglia, Norwich Research Park, Norwich NR4 7TJ, UK
e-mail: a.korobkin@uea.ac.uk

S. Malenica
Bureau Veritas, Research Department, Defense, Paris, France
e-mail: sime.malenica@bureauveritas.com

P. J. Aston (ed.), *More UK Success Stories in Industrial Mathematics*,
Mathematics in Industry 42, https://doi.org/10.1007/978-3-031-48683-8_8

known as whipping. Both slamming and whipping are highly perilous for ship safety and can cause major structural damage and even lead to vessel breakup. On the 18th January 2007 the 62,000 tonne container ship MSC Napoli was lost in heavy seas in the English Channel. Catastrophic damage to the ship hull caused by wave slamming led to a £120,000,000 clean-up operation lasting two and a half years. The official disaster report [1] was released documenting the catastrophic effect of wave slamming. It was stated in particular "In view of the potential increase in wave loading due to whipping effect, further research is required by classification societies to ensure that the effect is adequately accounted for in ship design and structural analyses, and that sufficient allowance is made for the effect when determining design margins."

Economic loss is not the most serious issue when large ships capsize. In 1994 the ferry MS Estonia capsized in the Baltic Sea when a slamming event caused its bow ramp to fail, allowing water into the vehicle deck, the ship to sink and 852 lives to be lost.

An expert knowledge of the dynamics involved in such violent events is crucial for the design of ships and offshore structures. This includes knowledge of the forces generated during slamming and of the stresses created throughout the structure of the ship. It requires a deep understanding of the highly complex fluid-structure interaction between the ship and the water. The loads involved are so massive that even apparently rigid structures deform elastically. Design processes to mitigate the risk of damage require resource-intensive CFD modelling for the fluid motion, and finite-element codes for the structural response of the ship. But these approaches must be guided by a fundamental understanding of the dynamics of the problem obtained through mathematical modelling. To add to the challenge, a computational interface module must be carefully designed to allow the standalone CFD and finite-element codes to communicate. Interface design remains a highly challenging task for which deep mathematical understanding of the whole problem is essential. Theoretically speaking such an interface should be constructed very carefully for slamming simulations, where added-mass effects play the leading role, see Sect. 2. On the other hand, different interfaces are used by different companies on the daily basis. The authors are unaware of interfaces, which are fully theoretically justified. All existing interfaces are made in-house and use different ways of iterations between a CFD code and a finite-element structural code. Iterations in the problems of hydroelastic slamming mean that, at the beginning of each time step, the shape and velocity of the structural surface are taken from the previous time step and applied as an initial guess, then a CFD code provides hydrodynamic loads acting on the structure. The calculated loads are applied to the finite-element code, which provides new shape and velocity of the structure. Then we return to the CFD code, calculate new hydrodynamic loads, and then update shape and speed of the structure once again. Repeating these steps several time, we may expect to converge to certain loads and structural response, which are considered as the actual coupled response. Then we proceed to the next time step, and repeat iterations again. Different existing interfaces use different iteration schemes and stop criteria. To achieve the convergence of iterations, under-relaxation parameter(s) can be used. This algorithm with under-relaxation works well in some cases and fails in others. A success depends on experience of a researcher.

A coupling interface between a CFD code and a structural code is a crucial element of hydroelastic slamming simulations. However, it seems that there is no a rational background of interface design. Computations are treated as a success, if they converge to something, which potentially could be wrong. It is shown in Sect. 3 that a proper coupling of hydrodynamic loads and structural response may increase the maximum strain in the structure by 20%, which is a lot, but also decrease the response in the impulsive regime. To arrive at a reliable risk assessment of naval structures, one needs to justify the interface code and validate the simulations using semi-analytical solutions in particular simplified situations.

2 Simplest Coupled Problem of Fluid-Structure Interaction

In this section, we consider a simple problem of fluid-structure interaction (FSI) to illustrate performance of an interface between a hydrodynamic code and a structural code. The problem is depicted in Fig. 1. A rigid sphere of radius R and mass M is submerged in an unbounded fluid and is supported by a spring of rigidity k. The fluid is incompressible and inviscid. There is no gravity and any other mass forces in this problem. Initially the sphere is displaced by a small distance x_0 from its equilibrium position and then released with zero speed. The motion of the sphere is one-dimensional. We shall determine the time-dependant displacement $x(t)$ of the sphere, where t is the time and $x(0) = x_0$, $x'(0) = 0$ at $t = 0$. This problem is a FSI problem but it does not represent problems of hydroelastic slamming. However, it contains the element of the hydroelastic slamming problems of our main concern, which is the coupling between the motion of the structure and the hydrodynamic loads under dominance of the inertia effects. The motion of the sphere depends on the hydrodynamic loads, which in turn depend on the sphere motion.

The motion of the body is governed by the equation

$$M\frac{\mathrm{d}^2x}{\mathrm{d}t^2} + kx = F_h\left(t, x(t), \frac{\mathrm{d}x}{\mathrm{d}t}, \frac{\mathrm{d}^2x}{\mathrm{d}t^2}\right), \tag{1}$$

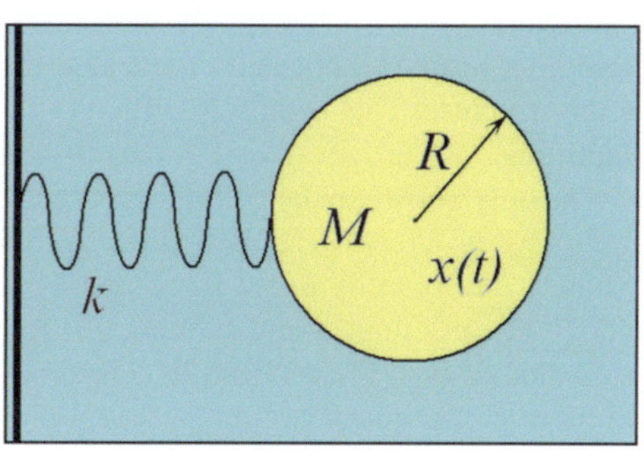

Fig. 1 A sphere of radius R and mass M in unbounded inviscid fluid supported by a spring of rigidity k

where F_h is the hydrodynamic force acting on the body, $dx/dt = x'(t)$ is the body velocity, and d^2x/dt^2 is the body acceleration. Equation (1) is to be integrated in time subject to the initial conditions $x'(0) = 0$ and $x(0) = x_0$. The theory of ordinary differential equations requires the Eq. (1) to be resolved with respect to its highest order derivative, d^2x/dt^2, and integrated in time thereafter. Therefore, Eq. (1) should be treated first as an equation with respect to the body acceleration $x''(t)$. The dependence of the hydrodynamic force on the body acceleration makes Eq. (1) complicated.

The hydrodynamic force F_h can be calculated by CFD solving fully nonlinear Navier-Stokes equations for certain position, velocity and acceleration of the body at each instant of the motion. However, within the present potential flow model of unbounded and inviscid fluid, it is known that the force is given by

$$F_h = -M_a \frac{d^2x}{dt^2}, \qquad M_a = \frac{2}{3}\pi \rho R^3, \tag{2}$$

where ρ is the fluid density and M_a is the added mass of the sphere in the unbounded fluid. Equation (1) represents the structural part and Eq. (2) represents the hydrodynamic part of the coupled FSI problem. Substituting Eq. (2) in Eq. (1) and collecting the terms with the body acceleration, we arrive at the well known result that a one-dimensional motion of a body in an unbounded fluid is the same as the motion of the body without the fluid but with an increased mass $M + M_a$.

Theoretically the problem Eqs. (1) and (2) is easy to solve in full details. However, the problem becomes very complicated if the hydrodynamic force is not given by a formula but is obtained by CFD. In this case, we can try to solve Eq. (1) with respect to x'' by iterations, as it is described in the Introduction. Having in mind that Eq. (2) should well approximate the CFD force for the given conditions, we may conclude that these straightforward iterations converge only if the added mass M_a is smaller that the structural mass M. For example, an official table ping pong ball is of mass $M = 2.7$ gram and of radius $R = 2$ cm. The added mass of the ball in water is $M_a = 16.7$ gram. Then $M_a/M = 6.24$ and iterations applied to Eq. (1) do not converge. On the other hand, the added mass of the ball in air is $M_a = 0.02$ gram and the iterations converge. The problem of Eq. (1) with the hydrodynamic force obtained by a CFD code can be solved by iterations between a CFD code and a structural code for the sphere motion in air but not in water. In the hydroelastic slamming problems normally the added mass of a structure is greater than the mass of the structure. This is the key difficulty with numerical simulations of hydroelastic slamming.

The only known to the authors approach of resolving this difficulty in hydroelastic slamming problems was suggested by Sun and Faltinsen [2]. They suggested to evaluate the hydrodynamic force approximately within the so-called von Karman model of water impact and subtract this force from both the CFD force on the right-hand side of Eq. (1) and from its left-hand side. This approach being applied to the problem of the sphere motion provides the following equation

$$(M + M_a)\frac{d^2x}{dt^2} + kx = F_{h,CFD}\left(t, x(t), \frac{dx}{dt}, \frac{d^2x}{dt^2}\right) + M_a\frac{d^2x}{dt^2}, \tag{3}$$

where $F_{h,CFD}$ stands to the hydrodynamic force computed by CFD. The right-hand side of Eq. (3) is not zero, but its dependence on the body acceleration is small compared with the first term on the left-hand side of the equation. Equation (3) is suitable for solving it by iterations with respect to the body acceleration.

3 Dynamic Amplification Factor in Hydroelastic Slamming

Problems of hydroelastic slamming can be solved as decoupled problems even, formally speaking, this is not right. However, the decoupled approach is much simpler and potentially can provide reasonable results. Within the decoupled approach the hydrodynamic loads acting on a structure are determined without account for the structural deformations and assuming the structure is rigid. Then these loads are applied to the elastic structure to determine stresses in the structure and assess the risk to the structure damage. It is important that the hydrodynamic loads and the structural response can be computed by dedicated solvers without an interface between them. In practice, even simpler approach is used, where the computed hydrodynamic loads are applied to the structure without account for its inertia. It is helpful to know how the levels of the stresses computed by the coupled and simplified quasi-static decoupled approaches are related. Such a measure is known as a dynamic amplification factor (DAF), which can be defined in different ways.

It is known that in hydroelastic slamming the hydrodynamic loads are high in magnitude but localised in time and space with extreme pressure peaks being ultimately filtered by the structural dynamics. If applied directly on the structure in quasi static sense, these extreme pressures would induce severe structural failure in many conditions. This means that the full hydroelastic coupling is absolutely necessary for the evaluation of the structural response. On the other hand, the idea of the *equivalent static slamming pressure* (ESSP) and pre-calculated DAF is proposed by some classification societies for direct strength assessment and scantling determination, see [2]. The definitions of ESSP by different classification societies are different, and their technical backgrounds are unclear. The DAF for a structure can be determined by the approach suggested in [3] and developed in [4, 5]. The DAF for elastic wedge entering water at constant speed is shown in Fig. 2. Here the DAF is defined as the maximum strain in the wedge plating computed by a fully coupled approach divided by the maximum strain calculated by the decoupled quasi-static approach. The horizontal axis is for the ratio of the impact time t_w, which is the duration of the impact loads before the wedge is completely wetted, to the period of the lowest wet mode of the wedge plating t_1. The solid line in Fig. 2 is obtained by solving the two-dimensional coupled problem within the Wagner model of water impact using the normal mode method. The DAF calculated by the simplified model well approximates the numerical results shown by markers. It is seen that dynamic strains can

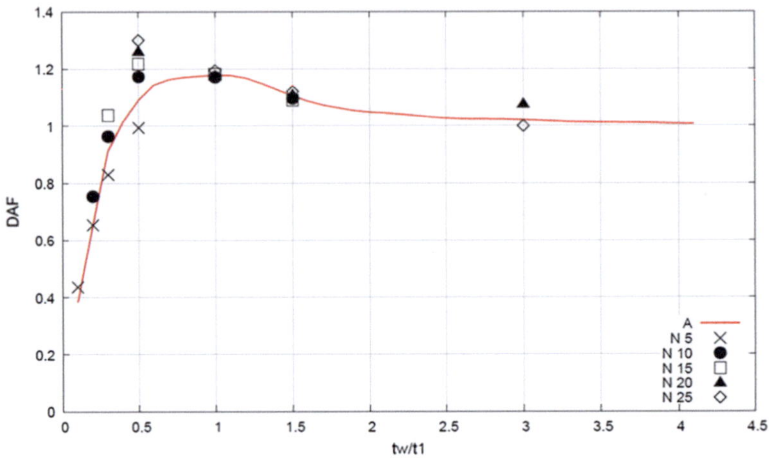

Fig. 2 Dynamic amplification factor for elastic wedge entering water at constant speed. The markers show the CFD results for different deadrise angles of the wedge. The solid line is for the DAF by the Wagner approach. The impulsive regime is for small t_w, where DAF< 1, dynamic one for moderate t_w, DAF> 1 and quasi-static regime for large t_w, DAF≈ 1

be 20% higher than the corresponding quasi-static strains and several times smaller in the impulsive regime.

4 Conclusion

Slamming represents a very important practical problem in the design of floating structures. Due to its sensitivity to the local flow features, extremely high local pressures both from local and global points of view might occur for some conditions. In those cases the numerical modelling of hydro structure interactions becomes very complex. The fully coupled hydroelastic models are required and they need to be validated very carefully. This should be done for all possible interaction regimes (impulsive, dynamic and quasi static). The use of the equivalent design pressure concept should be considered with care. This is especially true for the impulsive regime, where in spite of large pressure peaks, their effect on the structural response is not necessarily critical. This is because the large pressures are very localised in space and in time so that they are "filtered" by the structural dynamics. The global effects of slamming, known as whipping, should also be considered with care and fully coupled hydroelastic model need to be used.

References

1. Report on the investigation of the structural failure of MSC Napoli English Channel on 18 January 2007, Marine Accident Investigation Branch, Report No 9/2008, April 2008.
2. Sun, H., Faltinsen, O. M. (2006). Water impact of horizontal circular cylinders and cylindrical shells. Applied Ocean Research, 28(5), 299–311.
3. Zhu, L., Duan, L., Chen, M., Yu, T. X., Pedersen, P. T. (2020). Equivalent design pressure for ship plates subjected to moving slamming impact loads. Marine Structures, 71, 102741.
4. Khabakhpasheva, T. I., Korobkin, A. A. (2013). Elastic wedge impact onto a liquid surface: Wagner's solution and approximate models. Journal of Fluids and Structures, 36, 32–49.
5. Korobkin, A.A., Khabakhpasheva, T.I., Malenica, S. (2021) Hydrodynamic pressure in hydroelastic impact problems. In: Proceedings of the 36th International Workshop on Water Waves and Floating Bodies, 25–28 April 2021.
6. Malenica S., Gatin I., Seng S., Jagite G., Diebold L., Khabakhpasheva T., Korobkin A.A., (2021) Some aspects of the local hydro-structure interactions during hydrodynamic impacts. In: Proceedings of the 23rd Numerical Towing Tank Symposium (NUTTS), Duisburg, Germany.

Rolls-Royce Hydra CFD Code for Gas Turbine Engine Design

Mike Giles and Leigh Lapworth

Abstract In 2012, Hydra became Rolls-Royce's primary computational fluid dynamics (CFD) code for the aerothermal design of gas turbine engines used worldwide to power large commercial aircraft and for other purposes including power generation and naval propulsion. This article describes the history of its initial development at the University of Oxford, outlines some of its key technical features and the underlying mathematics, and discusses some of the engines developed using it. With its impact continuing today, Hydra was the basis of REF (Research Excellence Framework) impact case studies in both 2014 and 2021.

1 History of Development

Rolls-Royce plc is one of the UK's premier engineering companies. It operates worldwide and its gas turbine engines power around 50% of the world's large commercial aircraft. It has around 41,000 permanent employees worldwide, more than half in the UK, and its annual, revenue in 2023 was £15.5B.

When the first author joined the company in 1978 as an Undergraduate Engineering Apprentice, before studying mathematics at Cambridge University, the aerothermal design process for new engines was largely experimental, based on rescaling components from previous successful designs and guided by simple one-dimensional models of the averaged flow conditions along the engine. Today, it is almost completely computational, with detailed simulations and optimisation of structural stresses, deformations and vibrations; combustion; heat transfer; aerodynamics. Collectively, the computational design system greatly improves the quality of the final design, and at the same time greatly speeds up the design process.

M. Giles (✉)
Mathematical Institute, University of Oxford, Woodstock Road, Oxford OX2 6GG, UK
e-mail: mike.giles@maths.ox.ac.uk

L. Lapworth
Rolls-Royce plc, Box 31 Moor Lane, Derby DE24 8BJ, UK
e-mail: leigh.lapworth@rolls-royce.com

P. J. Aston (ed.), *More UK Success Stories in Industrial Mathematics*,
Mathematics in Industry 42, https://doi.org/10.1007/978-3-031-48683-8_9

Since 2012, Hydra has been Rolls-Royce's primary aerothermal analysis and design code. Its initial development was carried out by the first author and his research team at the University of Oxford, with the second author responsible for the project oversight in the early stages and then the continued development once it moved into Rolls-Royce.

The origins of the research go back to 1992 when Rolls-Royce facilitated the first author's move from MIT to Oxford and he established the Rolls-Royce University Technology Centre (UTC) in CFD, one of several UTCs which Rolls-Royce supports in a variety of different technical areas.

After a number of projects, including research on linearised flutter and forced response analysis, and the development of the OPlus library to support very large parallel CFD computations on unstructured grids, the initial development of Hydra took place during the period 1998–2006, with 6 PhD students and PDRAs (Post-Doctoral Research Associates) involved at different stages. Responsibility for further development of the software then transferred to a dedicated CFD Methods team in Rolls-Royce, with other Rolls-Royce UTCs contributing some new features.

New research involving two PDRAs began at the University of Oxford during 2009–2016 to enhance the parallel computing features of the code, enabling the use of large numbers of both multicore CPUs and GPUs, essential for carrying out the largest calculations. Further research and development on this aspect of the work continues in the Department of Computer Science at the University of Warwick.

2 Technical Details

Hydra is a finite volume CFD code, and its main features are

- a very flexible computational grid structure with a general multigrid formulation for efficiency;
- a variety of operating modes: steady, unsteady, linearised harmonic, adjoint;
- a parallel computing implementation providing the ability to perform extremely large computations.

A key requirement for Hydra was the ability to analyse the flow through very complex turbomachinery components. For this reason it works with very general grids with a mixture of cell types and then uses a finite volume discretisation to convert the unsteady compressible Navier-Stokes equations into a set of coupled nonlinear ODEs of the form

$$\frac{\mathrm{d}U}{\mathrm{d}t} + N(U) = 0,$$

where U is a vector consisting of the flow variables at all of the grid points, and $N(U)$ is a sparse nonlinear flux operator based on the fluxes along edges between neighbouring grid points.

Marching this to steady-state, with acceleration provided by a multigrid method, gives a steady flow solution. Alternatively, the unsteady equations can be solved using an implicit time-marching method which again uses multigrid to solve the implicit set of equations at each timestep. Later developments at the Surrey UTC added a fully implicit solver.

Flutter analysis requires the calculation of the unsteady forces on each blade due to their motion. Linearising the equations and considering a single harmonic which is the real part of $\exp(i\omega t)\,\widehat{U}$ leads to the linear harmonic equations

$$i\omega\widehat{U} + A\,\widehat{U} = \widehat{f},$$

where the driving term $\widehat{f}$ is due to the oscillation in the blades and the consequent motion of the computational grid. A similar form of equation is obtained for linearised forced response analysis, determining the forces on the blades due to the flow unsteadiness caused by the neighbouring rotating blade row; in this case $\widehat{f}$ comes from the unsteady boundary conditions. In both cases the linear harmonic equations are again solved using the multigrid.

An important theoretical concern with this approach is whether it is correct to consider infinitesimal perturbations of shocks which are almost discontinuous. The validity of this was determined in prior research at MIT [1] which showed that it is valid provided the numerical approximation of the shock is spread over a small number of points due to numerical smoothing. Another important implementation aspect concerns the generation of the code to compute $A\,\widehat{U}$. Manual differentiation is possible, of course, but it is tedious and error-prone and would need to be re-done each time the discretisation $N(U)$ is modified. Instead, Automatic Differentiation (AD) tools were used to automatically generate the code to compute $A\,\widehat{U}$ given the code to compute $N(U)$.

With design optimisation, the frequency is zero and the linear perturbation equations are

$$A\,\widetilde{U} = \widetilde{f},$$

with different values for $\widetilde{f}$ depending on the shape of the perturbation to the geometry. The problem is that there may be as many as 1000 or more different design perturbations. However, there is often just one output quantity of interest, such as some measure of the efficiency, and after linearising perturbations to this, the goal is to evaluate the product $g^T\widetilde{U}$ for each one of the different $\widetilde{f}$ forcing terms. Defining the adjoint variables V as the solution to

$$A^T V = g,$$

then

$$g^T\widetilde{U} = V^T A\,\widetilde{U} = V^T\widetilde{f},$$

so a single calculation of the adjoint variables V followed by a set of simple dot-products with each $\widetilde{f}$ gives the desired outputs [2]. This is the adjoint capability

Fig. 1 Mach number contours for a 4.5 billion grid point simulation of a German Aerospace Center (DLR) 4.5 stage research compressor

within Hydra and its development required research into an efficient adjoint of the standard multigrid solver to ensure an identical rate of convergence to the multigrid solver applied to solve $A\,\widetilde{U} = \widetilde{f}$ [3]. Again AD tools were used to automatically generate the code to compute $A^T V$; doing this manually would have been extremely time-consuming and very hard to debug.

Parallel scientific computing lies at the interface between mathematics and computer science. It requires computer science knowledge and expertise to obtain high performance on the latest HPC architectures with many compute nodes, each of which may have multiple multicore CPUs and/or one or more GPU accelerators. However, a thorough understanding of the mathematical structure of the application, and others like it, is essential to formulate an appropriate *abstraction*, separating out *what* needs to be computed from *how* it is computed. In the case of Hydra, at an abstract level the algorithm requires a loop over the *edges* of the computational grid, performing various flux calculations which can be done in parallel provided two flux calculations do not try to update the same grid node at the same time.

The first Hydra implementation used a previously developed parallel library OPlus [4] with Hydra specifying these parallel loops and the data dependencies, and the OPlus library automatically partitioning the problem across multiple single-core computers using MPI message-passing to coordinate the parallel execution. In more recent research [5] an updated framework OP2 has been developed with the same fundamental abstraction but using automated code generation to create optimised code for compute nodes with either multicore CPUs or many-core GPUs. This enables huge calculations, the largest of which to date had 4.5 billion grid points, and executed on a total of 65,536 CPU cores spread across 512 compute nodes of the UK's national supercomputer ARCHER2 [6]. Figure 1 shows axial velocity contours from this calculation.

3 Use Within Rolls-Royce

Simulation and modelling, enabled by high performance computing, have long been indispensable design tools. Moreover, competitive products require competitive tools and this is exactly what Hydra provides to Rolls-Royce. The technology developed for the Trent XWB, the world's most efficient large aero-engine, could only have been achieved through simulation and modelling. Collectively, these delivered a 15% improvement in fuel consumption relative to the original Trent engine and provide $2.9M savings per year per aircraft in fuel alone. There are many aspects to designing a gas turbine engine but one of the most important is aerodynamic performance. Hydra provided the 3D CFD simulations that resulted in the Trent XWB fan having world-beating levels of performance and noise, and the turbine system having the highest efficiency of any Trent engine.

With the first production release dating from around 2005, Hydra has been the corporate CFD code for over a decade and is used globally by engineers in the UK, US, Germany and Singapore. There are well over 1,000,000 runs of Hydra a year in the UK alone. Hydra has been used on a range of products around the globe including, for example, the Pearl 700 engine for the Gulfstream G700 ultra-long-range business jet. This engine has 3.5% less fuel burn and 5% greater efficiency than the previous BR725 engine for the G650.

In addition to production usage, Hydra provides the back-bone for the majority of Rolls-Royce's research into CFD methods and their applications. This has involved significant and ongoing research collaborations with a number of UK universities. These range from new modelling capabilities to performance enhancements for new and emerging hardware. It is a testament to the lead author's vision in designing the code that Hydra continues to have world class parallel performance. It has already achieved petascale performance and exascale is expected in the coming years.

In-house research has built on the adjoint capabilities of Hydra which provides the core CFD solver for the Company's design and optimisation suite. This is coupled to the Company's advanced meshing capabilities that allow, for example, high fidelity digital twins to simulate manufacturing variations derived from optical scans [7]. Figure 2 shows the digital twin for the bypass outlet guide vanes from a case study which demonstrated how significant cost savings for a fleet of engines could be made.

It has been over 20 years since Rolls-Royce installed the first prototype version of Hydra on an in-house system. New capabilities are continually being added and new research collaborations started - we fully expect to still be using it in 20 years' time and for it to still provide a world-class capability. Hydra has already contributed to the reduction in emissions and better fuel efficiency of Rolls-Royce products. It will also be vital in Rolls-Royce's ambition to play a fundamental role in enabling the sectors in which it operates to achieve net zero carbon by 2050.

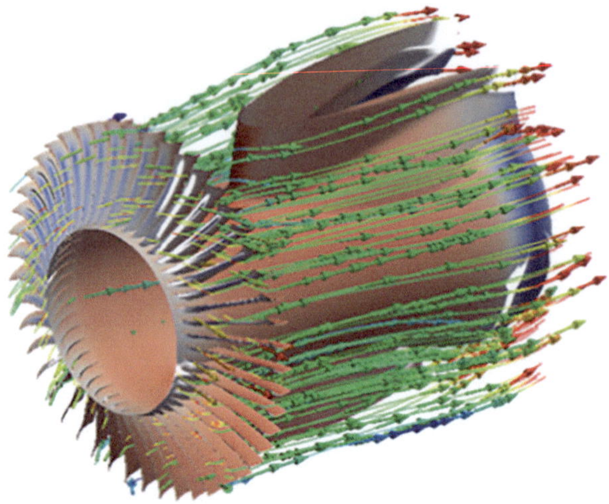

Fig. 2 Surface static pressure contours and velocity vectors from a Hydra simulation of the bypass duct [7]

Acknowledgements The contributions of many students and postdoctoral research associates are acknowledged, including Sergio Campobasso, Paul Crumpton, Mihai Duta, Devendra Ghate, Dana Lindquist, Pierre Moinier, Gihan Mudalige, Jens-Dominik Müller, Niles Pierce, and Istvan Reguly. The contributions of the Rolls-Royce in-house developers are acknowledged including Paolo Adami, Richard Bailey, David Radford, Indi Tristanto, Kurt Weber, Matthew West. In addition to substantial funding from Rolls-Royce, the research was supported by EPSRC through research grants "Computation of viscous flows using unstructured grids", GR/L18280/01, 1996–99; "Adjoint CFD algorithms for aeronautical design", GR/L18280/01, 1998–2001; "AD2 Compeng: Automatic differentiation and adjoints applied to computational engineering", GR/R85358/01, 2003–2006; "Multi-layered abstractions for PDEs", EP/I006079/1, 2010–2014.

References

1. Dana R Lindquist and Michael B Giles. Validity of linearized unsteady Euler equations with shock capturing. *AIAA Journal*, 32(1):46–53, 1994.
2. Michael B Giles and Niles A Pierce. An introduction to the adjoint approach to design. *Flow, Turbulence and Combustion*, 65(3):393–415, 2000.
3. Michael B Giles, Mihai C Duta, Jens-Dominik Muller, and Niles A Pierce. Algorithm developments for discrete adjoint methods. *AIAA Journal*, 41(2):198–205, 2003.
4. Paul I Crumpton and Michael B Giles. Multigrid aircraft computations using the OPlus parallel library. In *Parallel Computational Fluid Dynamics* 1995, pages 339–346. Elsevier, 1996.
5. István Z Reguly, Gihan R Mudalige, Carlo Bertolli, Michael B Giles, Adam Betts, Paul HJ Kelly, and David Radford. Acceleration of a full-scale industrial CFD application with OP2. *IEEE Transactions on Parallel and Distributed Systems*, 27(5):1265–1278, 2015.
6. Gihan R Mudalige, Istvan Z Reguly, Arun Prabhakar, Dario Amirante, Leigh Lapworth, and Stephen A Jarvis. Towards virtual certification of gas turbine engines with performance-portable simulations. In *2022 IEEE International Conference on Cluster Computing (CLUSTER)*, 2022.
7. Shahrokh Shahpar. Building digital twins to simulate manufacturing variation. In *Turbo Expo: Power for Land, Sea, and Air*, volume 84065, page V02AT32A049. American Society of Mechanical Engineers, 2020.

Multiparametric Dynamic Modelling of Lightweight Elastic Structures

D. Prikazchikov, J. Kaplunov, L. Prikazchikova, and G. A. Rogerson

Abstract Recent trends in the automotive industry prioritise the reduction of CO_2 emission by implementing advanced lightweight materials and structures within car production. An unwanted consequence of lightweight design is an increase in the level of vibration and noise. Keele's applied mathematicians established a powerful multi-parametric framework to aid dynamic modelling of strongly inhomogeneous structures involving lightweight components, e.g. sandwich plates. The developed approach is especially useful to elucidate the particularly harmful low-frequency vibration regimes. The methodology has been adopted by the TPV Group for the manufacturing of modern automotive components, both reducing production costs and promoting a more sustainable environment.

1 Introduction

The underpinning research has been inspired by the major industrial challenge designing lightweight parts for more environmentally friendly vehicles. This is currently in great demand due to the importance of CO_2 emission reduction. However, implementation of lightweight components may cause an unwanted increase in vibration and noise. Thus, the key feature of the complex underlying original engineering problem presents two mutually contradictory factors.

A robust methodology for the modelling of lightweight multi-layered and multi-component elastic structures was developed by Keele applied mathematicians. The

D. Prikazchikov (✉) · J. Kaplunov · L. Prikazchikova · G. A. Rogerson
School of Computer Science and Mathematics, Keele ST5 5BG, Staffordshire, UK
e-mail: d.prikazchikov@keele.ac.uk

J. Kaplunov
e-mail: j.kaplunov@keele.ac.uk

L. Prikazchikova
e-mail: l.prikazchikova@keele.ac.uk

G. A. Rogerson
e-mail: g.a.rogerson@keele.ac.uk

P. J. Aston (ed.), *More UK Success Stories in Industrial Mathematics*,
Mathematics in Industry 42, https://doi.org/10.1007/978-3-031-48683-8_10

focus was on the multi-parametric dynamic analysis of "lightweight" automotive components. The novel mathematical models involve multi-component (or multi-layered) high-contrast structures containing both stiff and soft parts. Breakthrough solutions have been achieved by studying a variety of configurations of composite materials and structural elements, aiming at an overall weight reduction.

The Keele group pioneered work on the multi-parametric dynamic analysis of strongly inhomogeneous elastic structures. Apart from the basic small parameter corresponding to the presence of lightweight components, additional parameters related to small/large ratio of stiffness and linear size of components, were taken into consideration. The resonant phenomena caused by the excitation of the most harmful low-frequency vibrations, due to high contrast, were rigorously investigated. A number of new insights were made, including a two-mode asymptotic theory for bending vibrations of a sandwich plate, replacing or refining the conventional single-mode asymptotic formulations or various ad hoc approximations.

It was noted in [1] that the natural high contrast in properties of a lightweight structure causes a shift of the first thickness shear resonance to the low frequency band in the case of plate bending, resulting in its interaction with the fundamental mode. This is in line with previous observations for strongly inhomogeneous sandwich plates [2] and also related to low-frequency vibrations of multi-span bars, see [3, 4]. This phenomenon of an additional low-frequency mode in layered structures is closely linked to a degenerating (slowly decaying) boundary layer in statics, see [5]. It also leads to considerable challenges in direct implementation of finite element computations and interpretation of the results, stimulating a far more elaborative theoretical insight.

2 Analysis of Low-Frequency Vibrations of Lightweight Structures

As an example, consider the elastodynamics of a lightweight elastic layered symmetric infinite sandwich laminate with a light core of thickness $2h_c$ and heavy skins of thickness h_s, see Fig. 1.
Here and below subscripts "c" and "s" are associated with the core and skin layers, respectively. Both skins and core are assumed to be linearly isotropic, with free outer faces of the laminate and perfect bonding on the interfaces between layers.

Fig. 1 A lightweight laminate

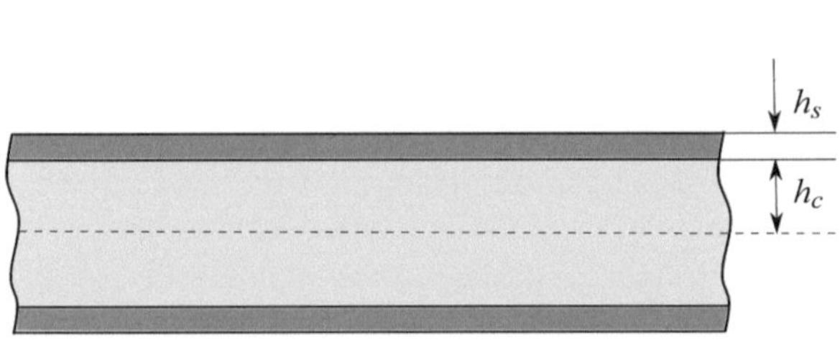

A natural small parameter is the ratio of volume mass densities

$$\rho = \frac{\rho_c}{\rho_s} \ll 1. \tag{1}$$

For the sake of definiteness we assume the ratio of shear moduli and inverse ratio of thickness satisfy

$$\mu = \frac{\mu_c}{\mu_s} = \mu_0 \rho, \qquad h = \frac{h_s}{h_c} = h_0 \rho^a, \tag{2}$$

where $\mu_0 \sim 1$, $h_0 \sim 1$, and the parameter $0 \le a < 1$ controls the proportion of light core within the structure. As we will see, its value is potentially important for structural optimisation, leading to either a uniform or composite type of asymptotic approximation in the low-frequency domain of interest.

Below the results of the two-parametric analysis in ρ and a of the full dispersion relation (antisymmetric case) of the considered elastic laminate are presented, focusing on the long-wave low-frequency limit, in which

$$K = kh_c \ll 1, \qquad \Omega = \omega h_c \sqrt{\frac{\rho_c}{\mu_c}} \ll 1, \tag{3}$$

where ω and k are angular frequency and wave number, respectively. Then, the approximation for the lowest shear cut-off frequency, i.e. the lowest non-zero solution of the dispersion relation at $K = 0$, may be found in the form

$$\Omega_{sh} \approx \sqrt{\frac{\rho}{h}} \sim \rho^{(1-a)/2} \ll 1, \tag{4}$$

belonging to the low-frequency zone, see [2], and also a closely related problem for a three-component rod [3].

The dispersion relation is expanded as Taylor series over the long-wave low-frequency domain (3), yielding

$$\gamma_1 \Omega^2 + \gamma_2 K^4 + \gamma_3 K^2 \Omega^2 + \gamma_4 K^6 + \gamma_5 \Omega^4 + ... = 0, \tag{5}$$

where γ_j are quantities depending on the material and geometric parameters of the layers, see [1] for more detail.

It may be shown that the value of the parameter $a = \frac{1}{3}$ is critical, with the range of $0 \le a \le \frac{1}{3}$ corresponding to the uniform two-mode low-frequency approximation

$$\rho^{1-a} G_1 \Omega^2 + \rho^{1-a} G_2 K^4 + G_3 K^2 \Omega^2 + \frac{1}{3} \rho^{2a} G_2 K^6 + G_5 \Omega^4 = 0, \tag{6}$$

whereas the range $\frac{1}{3} < a < 1$ is associated with a composite approximation

$$\rho^{1-a}G_1\Omega^2 + \rho^{1-a}G_2K^4 + G_3K^2\Omega^2 + G_5\Omega^4 = 0. \tag{7}$$

Here

$$G_1 = -h_0\mu_0^2, \quad G_2 = 4h_0\mu_0(1-\kappa_s^2), \quad G_3 = -h_0G_2, \quad G_5 = -h_0G_1, \tag{8}$$

where

$$\mu_0 = \frac{\mu}{\rho} \quad \text{and} \quad \kappa_s = \frac{c_{2s}}{c_{1s}}, \tag{9}$$

with c_{1s} and c_{2s} denoting the longitudinal and shear wave speeds within the skin layer, for more detail see [1]. For the composite approximation (7), it may be shown that in the vicinity of the lowest cut-off frequency the fundamental mode is already outside the long-wave region of interest, thus the approximation is likely to be valid over two non-intersecting frequency domains.

These results complement the previously noted scenarios of high-contrast three-layered elastic plates, considered in [2]. Numerical illustrations showing these two typical behaviours are shown in Figs. 2 and 3, where exact dispersion curves are shown in solid lines, with associated asymptotic approximations depicted by dashed lines. The plot on Fig. 2 illustrates the uniform asymptotic expansion (6), when $a = \frac{1}{9}$, whereas Fig. 3 shows a composite approximation (7), when $a = \frac{2}{3}$. A frequency gap is clearly seen in Fig. 3, with the long-wave approximation for the fundamental mode being invalid in the vicinity of the lowest cut-off frequency.

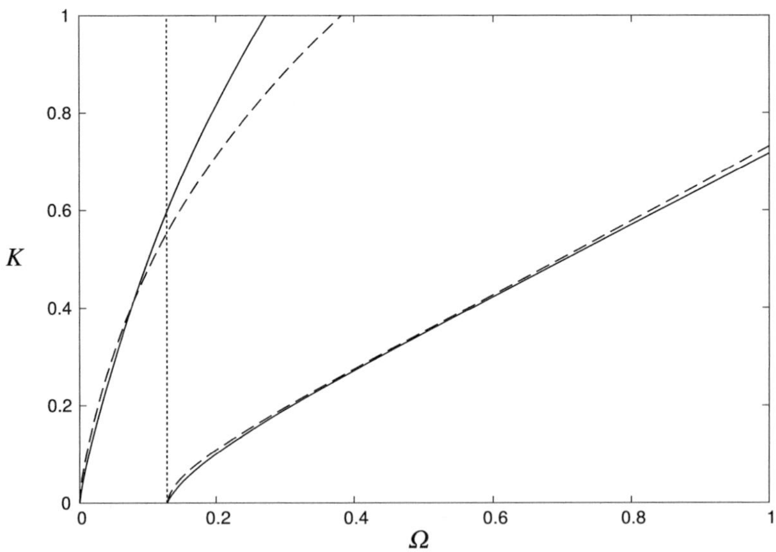

Fig. 2 Exact solution (solid line) and asymptotic approximation (6) (dashed line) for $a = \frac{1}{9}$

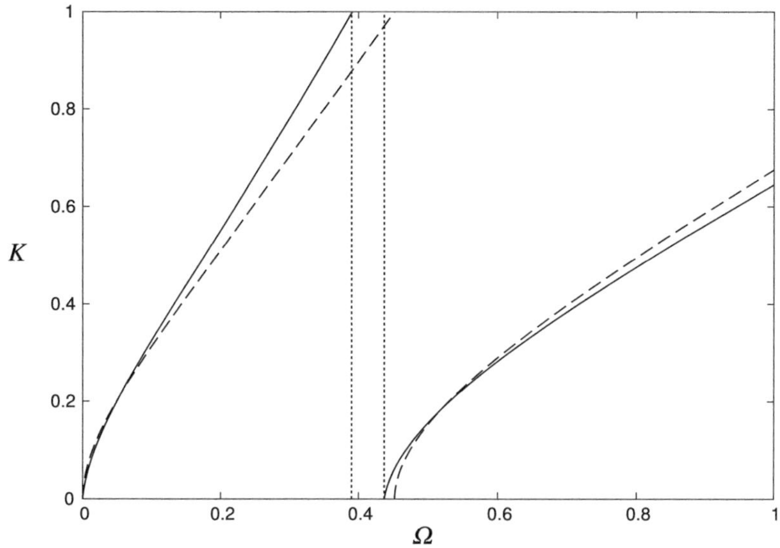

Fig. 3 Exact solution (solid line) and asymptotic approximation (7) (dashed line) for $a = \frac{2}{3}$

The presented methodology is not restricted to the setup of a three-layered symmetric plate. It can be extended to other types of lightweight laminates. In addition, extra problem parameters, including a strong contrast in the wave speed for core and skin layers, may be taken into consideration. Multilayered and multi-span structures can also be treated [4]. Finally, we mention a delicate formulation of boundary conditions for semi-infinite high contrast laminates, following on from the generalisation of the Saint-Venant's principle [5].

3 Details of Impact

The impact of the research conducted by Keele group originates from the industrial partnership within the EVA4Green EU project under the MOBILITY priority with the TPV Group, which is a large-scale company with headquarters in Slovenia, producing automotive parts for BMW, Daimler, Renault, Volkswagen, Volvo, and other manufacturers. The company has an annual turnover around €150 million, and over 1200 employees (2018). The MOBILITY priority area is identified as a key development within the international automotive industry, and EU industrial strategies.

Prior to this work, the ability to perform qualitative evaluation of the parameters characterising the dynamic behaviour in strongly inhomogeneous structures was rather limited. The crucial contribution of Keele applied mathematicians was showing that the resonant phenomena, typical of lightweight structural elements, involve additional low-frequency modes. In particular, for thin multi-layered plates, these vibrations interact with the conventional ones, leading to drastic changes in dynamic response. These studies resulted in a fairly universal but "user-friendly" methodology adopted by the R&D team at TPV Group, based on multi-parametric analysis of dynamic behaviour of strongly inhomogeneous elastic structures. The related semi-analytical procedure helps determine the optimal design of lightweight car parts and usually precedes 3D numerical finite element modelling, which is followed by experimentation, prototypes and manufacture.

The advanced mathematical methodology developed by the Keele research group, was the key element for:

(i) improved processes for designing lightweight car parts, in particular the cross-car beam and suspension arm, which are essential components for the stability, control, and safety system of a vehicle, and for which non-compromised design for lightweight construction presents substantial engineering challenges;

(ii) a lower carbon footprint, brought about both by the reduced amount of raw materials, especially steel, now needed for construction, and also by the reduction of energy consumption and harmful emissions from lighter vehicles;

(iii) a shorter development cycle, made possible because the scientific principles governing lightweight design are now accessible to TPV via the mathematical work of the Keele research group;

(iv) upskilling of the existing R&D staff at TPV, via a series of workshops led by the Keele mathematicians (with news disseminated in local media), enabling the R&D group at TPV to use the Keele models, analyses and software, which can only be implemented after special staff training; also translation [6] from English into Slovenian of the book "Dynamics of thin walled elastic bodies" by Kaplunov et al., previously published in Academic Press;

(v) creation of jobs through the securing from the EU of funding to Slovenia from ERDF and the European Cohesion Fund.

Overall, implementation of EVA4Green project has resulted in 22 new products whose competitive advantage within the market can be demonstrated by more than 40 innovations and 16 international patent applications. TPV Group's promotional material now specifically refers to their specialism in lightweight product design.

References

1. Kaplunov, J., Prikazchikov, D. A., Prikazchikova, L. A., Nikonov, A., Savšek, T.: Multi-parametric dynamic analysis of lightweight elastic laminates. In: IOP Conference Series: Materials Science and Engineering, **683(1)**, 012014 (2019).
2. Kaplunov, J., Prikazchikov, D. A., Prikazchikova, L. A.: Dispersion of elastic waves in a strongly inhomogeneous three-layered plate. Int. J. Solids Struct. **113**, 169–179 (2017).
3. Kaplunov, J., Prikazchikov, D., Sergushova, O.: Multi-parametric analysis of the lowest natural frequencies of strongly inhomogeneous elastic rods. J. Sound Vib. **366**, 264–276 (2016).

4. Kaplunov, J., Prikazchikov, D. A., Prikazchikova, L. A., Sergushova, O.: The lowest vibration spectra of multi-component structures with contrast material properties. J. Sound Vib. **445**), 132–147 (2019).
5. Kaplunov, J., Prikazchikova, L., Alkinidri, M.: Antiplane shear of an asymmetric sandwich plate. Cont. Mech. Thermodyn. **33(4)**, 1247–1262 (2021).
6. Kaplunov, J., Kossovich L., Nolde E.: *Dinamika tankostenskih elastičnih teles*. FINI Publisher, 2019, ISBN 978-961-94246-7-4.

Mathematical Modelling of Longitudinal Forces in Freight Trains Using the Concept of Almost Rigid Body Motion

J. Kaplunov, D. Prikazchikov, and L. Prikazchikova

Abstract Evaluation of longitudinal forces in freight trains is of primary importance for safe operation. To adequately predict the most harmful scenario, including the coupled impacts squeezing the cars, it is necessary to account for deformability, inhomogeneity, and energy dissipation. A robust mathematical model has been derived to incorporate the aforementioned phenomena. It was embedded into Amsted Rail Inc. software, directly impacting on the prediction of end-of car systems performance, calculation of the longitudinal forces in fluid-filled tanks, and implementation of accelerometers. The developed methodology had important applications in pre-sale customised modelling, optimisation of experimental setups, and patent activities.

1 Introduction

Prediction and assessment of the longitudinal forces between freight cars in heavily loaded long trains are crucial for maintaining operational safety and reducing the risk of accidents, such as train derailments or disruptions due to damage of cars, goods, and railway infrastructure, along with service losses. The drive by rail freight operating companies to increase the weight and length of trains, and at the same time reduce the shock experienced by the transported equipment and materials, requires a better understanding of the longitudinal forces involved.

When specifying the details of new rail freight cars, the operators supply technical characteristics of the train to be equipped, along with the peculiarities of the topographical profile, in order to proceed with customized design. The manufacturer designers then implement computer modelling for evaluating the longitudinal

J. Kaplunov (✉) · D. Prikazchikov · L. Prikazchikova
School of Computer Science and Mathematics, Keele University, Staffordshire ST5 5BG, UK
e-mail: j.kaplunov@keele.ac.uk

D. Prikazchikov
e-mail: d.prikazchikov@keele.ac.uk

L. Prikazchikova
e-mail: l.prikazchikova@keele.ac.uk

P. J. Aston (ed.), *More UK Success Stories in Industrial Mathematics*,
Mathematics in Industry 42, https://doi.org/10.1007/978-3-031-48683-8_11

forces arising in end-of-car systems, e.g. friction draft gears and hydraulic cushion units. Traditionally these methodologies are based mainly on the classical laws of rigid body dynamics and virtually ignore 'squeezing' of freight cars caused by high-level longitudinal forces typical for especially harmful coupled impacts. Internal dissipation of energy and inhomogeneity within carriages are also usually neglected.

The elementary approach to mathematical modelling of longitudinal motion of a freight train in terms of a chain of masses modelling carriages, connected by springs associated with draft gears, is common since N.E. Zhukovsky time (late XIX—early XX century). It should be noted that although the models of draft gears were becoming increasingly complex, in particular, involving nonlinearity, the carriages were usually modelled as masses concentrated in material points, with their motion governed by the Newton's Second Law for quite a while.

The challenges in modern railway industry motivate more accurate modelling of carriages, since the mass-spring model cannot account for a number of more sophisticated phenomena, including internal inhomogeneity, energy dissipation due to deformations of carriages, and also do not allow studying the scenario of self-equilibrated loads, related to coupled impact effect.

One of the hypothetical possibilities is related to full 3D modelling of a carriage embedding all its components into FEM software. However, at the moment the computational efforts required in case of long freight trains are considerably overwhelming the available capacities. Moreover, determination of exact values of the numerous parameters is also hardly possible nowadays.

The proposed approach relies on the approximate model, accounting for the effects of inhomogeneity and viscoelasticity of carriages and allowing treatment of self-equilibrated loading. Each carriage is modelled as a rod, with forces applied at the ends. A low-frequency perturbation of the boundary value problem for the wave equation results in an approximate integro-differential formulation for nearly rigid body motion of the rod, see [1, 2]. The latter takes the form of the generalised Newton's Second Law with extra terms taking into consideration the effect of viscoelastic microstructure, complementing the previous results in the field of dynamics of micro-inhomogeneous solids, e.g. see [3].

2 Almost Rigid Body Motion

Consider longitudinal vibrations of a viscoelastic inhomogeneous rod of length $2l$, subject to end forces F_1 and F_2, see Fig. 1. The 1D equation of motion is written as

$$m(x)u_{tt} = F_x, \tag{1}$$

where $m(x)$ is mass per unit length, $u = u(x, t)$ denotes horizontal displacement, $F = F(x, t)$ stands for the longitudinal force, and a subscript x or t indicates differentiation along the longitudinal coordinate x or time t as appropriate.

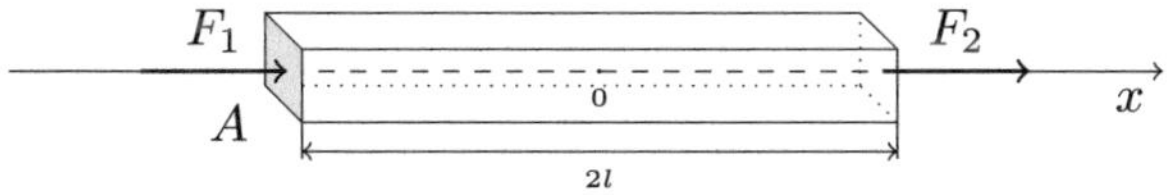

Fig. 1 Problem statement

Linear viscoelastic constitutive behaviour is taken as

$$u_x = \frac{1}{E_A}\left(F(x,t) - \int_0^t K\left(\gamma(x)(t-t_1)\right)\frac{\partial F(x,t_1)}{\partial t_1}dt_1\right), \tag{2}$$

with $E_A(x) = E(x)A(x)$, where $E(x)$ is Young's modulus, $A(x)$ is cross-sectional area, and the kernel $K(\gamma(x)t)$ depending on the function $\gamma(x)$, characterising local creep scale at the point with coordinate x.

The boundary conditions are taken as

$$F(-l,t) = F_1(t), \qquad F(l,t) = F_2(t), \tag{3}$$

see Fig. 1.

The focus below is on low-frequency motions described by the governing equations (1)–(3). Let us introduce the typical values of the variable quantities $m(x)$, $E_A(x)$ and $\gamma(x)$ as m_0, E_{A0} and γ_0, respectively. In this case, a typical time scale for viscous behaviour γ_0^{-1} is assumed to be much greater than the time taken by the longitudinal wave to propagate the distance between the ends of the rod, i.e.

$$\gamma_0^{-1} \gg l\sqrt{\frac{m_0}{E_{A0}}}. \tag{4}$$

The solution of the problem (1)–(3) under the assumption (4) can be obtained in the form of asymptotic series in terms of the small parameter

$$\varepsilon = \frac{l^2\gamma_0^2 m_0}{E_{A0}^2} \ll 1, \tag{5}$$

operating in scaled variables

$$\xi = \frac{x}{l}, \qquad \tau = \gamma_0 t, \tag{6}$$

along with the dimensionless quantities

$$u_* = \frac{u}{l}, \qquad F_* = \frac{F}{\varepsilon E_{A0}}. \tag{7}$$

Expanding the latter as asymptotic series

$$u_* = u_0 + \varepsilon u_1 + \ldots, \qquad F_* = f_0 + \varepsilon f_1 + \ldots, \tag{8}$$

we have at leading order

$$m_*(\xi) u_{0\tau\tau} = f_{0\xi}, \qquad u_{0\xi} = 0, \tag{9}$$

subject to

$$f_0(-1, \tau) = F_{1*}(\tau), \qquad f_0(1, \tau) = F_{2*}(\tau), \tag{10}$$

where

$$m_*(\xi) = \frac{m(\xi)}{m_0}, \qquad F_{i*} = \frac{F_i}{\varepsilon E_{A0}}, \quad i = 1, 2, \tag{11}$$

and, similarly to previous conventions, the subscript of ξ or τ indicates differentiation with respect to the appropriate variable.

It immediately follows from the second equation in (9) that $u_0(\xi, \tau) = v_0(\tau)$. Thus, the proposed approach is associated with almost rigid body behaviour. The solvability of the leading order problem above dictates

$$M_* \, v_{0\tau\tau} = F_{2*} - F_{1*}, \tag{12}$$

which might be recognised as the Newton's Second Law for a material point with the overall mass $M_* = \int\limits_{-1}^{1} m_*(\xi) d\xi$. The solution for f_0 can then be expressed as

$$f_0 = \frac{F_{2*} - F_{1*}}{M_*} \int\limits_{-1}^{\xi} m_*(\xi) d\xi + F_{1*}. \tag{13}$$

At next order, we have

$$m_*(\xi) u_{1\tau\tau} = f_{1\xi}, \quad u_{1\xi} = \frac{1}{E_{A*}} \left(f_0 - \int\limits_0^{\tau} K\left(\gamma_*(\xi)(\tau - \tau_1)\right) \frac{\partial f_0(\xi, \tau_1)}{\partial \tau_1} \, d\tau_1 \right), \tag{14}$$

where

$$E_{A*} = \frac{E_A(\xi)}{E_{A0}}, \qquad \gamma_* = \frac{\gamma(\xi)}{\gamma_0}, \tag{15}$$

along with the boundary conditions $f_1(\pm 1, \tau) = 0$.

Finally, we arrive at the refined formulation for the geometrical center of the rod, written in terms of original variables as

$$Ma = R - (1 - K(0)) \int_{-l}^{l} m(x) \int_{0}^{x} \frac{F_{0tt}(x_1, t)}{E_A(x_1)} dx_1 \, dx - K_t(0) \int_{-l}^{l} m(x) \int_{0}^{x} \frac{F_{0t}(x_1, t)}{E_A(x_1)} dx_1 \, dx$$
$$- \int_{-l}^{l} m(x) \int_{0}^{x} \frac{1}{E_A(x_1)} \int_{0}^{t} K_{tt}\big(\gamma(x_1)(t - t_1)\big) F_{0t_1}(x_1, t_1) \, dt_1 \, dx_1 \, dx, \qquad (16)$$

where $R = F_2 - F_1$, $a = u_{0tt} + \varepsilon u_{1tt}$ and $M = m_0 M_*$ are resultant force, acceleration and mass, respectively, and also $F_0 = \varepsilon E_{A0} f_0$.

The derived Eq. (16) governs almost rigid body motion of the rod, involving a low-frequency correction to the classical equation $Ma = R$, accounting for viscosity and inhomogeneity. This refinement also allows evaluation of the dynamic response caused by self-equilibrated external loads ($F_2 = F_1$).

3 Details of Impact

The main impact was achieved as part of industrial collaboration between Keele applied mathematicians and Amsted Rail Inc., USA, which is one of the largest and longest running companies in the field of railway engineering, having presence in all six continents (with annual turnover of around \$ 4B). To address the problem of prediction and assessment of the longitudinal forces between freight cars in heavily loaded long trains research efforts of Keele's group have been supported by two industrial contracts from Amsted Rail, in particular, funding a PhD studentship and a postdoctoral fellow.

Working with Amsted Rail, the Keele group has developed and implemented the concept of 'almost rigid body motion' presented in the previous section, see also [1, 2], developed in collaboration with Amsted R & T staff. This enables the identification and analysis of the most challenging regimes requiring consideration of the freight car deformability, inhomogeneity, and associated energy losses. Importantly, the model developed takes about the same computation time, but is efficient in a substantially broader range of operational scenarios compared to the routine procedures within the classical framework of rigid body dynamics.

Nowadays, the methodology developed by Keele group, combined with AI approach implemented in the Ph.D. thesis [4] (arising from the PhD studentship funded by Amsted), is integrated into the Amsted Rail software. This helps Amsted Rail to sustain its mission not only as a world leading provider and manufacturer, but also as a highly competent problem-solver able to apply its know-how on major international markets such as Australia, Brazil, China, India, and South Africa. A number of Amsted R & T projects involving computational modelling of specific train configurations all over the world have been already impacted by Keele's insights. It is remarkable that in critical regimes, including the so-called coupled impacts, a visible deviation from the conventional setup corresponding to rigid body dynamics is observed.

As a further development from the approach established by Keele's applied mathematicians, Amsted Rail experimental series for evaluating coupled car impacts were optimised. In particular, the approach overcame fundamental issues of high sensitivity of the experimental data read from the accelerometers to their location, by taking into consideration car deformability and inhomogeneity, as well as viscous properties of freight inside the car. It also improved the framework for neural network training by incorporating extra problem parameters. The developments in measuring of longitudinal forces have been patented by the company, with the reference to analytical approach.

It is also worth noting that while working on the industrial project with Amsted Rail [4], the PhD student managed to gain a high level of competence in the modern research methodologies widely used in Railway Engineering, including the AI systems (neural networks). Dr Shestakova was then employed by GHD, a world leading professional services company operating in the global market sectors, including transportation. There, her advanced modelling skills have substantially impacted on several major projects.

In addition, the scope of the proposed research has been broadened by incorporating the robust low-frequency theory for thin elastic shells, see [5]. This generalisation gives the R & T staff at Amsted Rail a powerful technique for mathematical modelling of railway fluid-filled tanks. To the best of our knowledge, there was no consistent theoretical methodology for evaluating longitudinal forces in fluid-filled tanks prior to recent developments of Keele applied mathematicians, obtained within the postdoctoral fellowship sponsored by Amsted Rail.

The impact of Keele research is not restricted to the contributions to Amsted Rail activities. Other railway companies, including Hum Industrial Technology have also benefited from joint efforts with Keele. Finally, it is worth noting that the research developments [1, 6], related to the refinement of rigid body dynamics are of fundamental nature, potentially feeding into more applications, including e.g. soft robotics.

References

1. Kaplunov, J., Shestakova, A., Aleynikov, I., Hopkins, B., Talonov, A.: Low-frequency perturbations of rigid body motions of a viscoelastic inhomogeneous bar. Mech. Time-Depend. Mater. **19(2)**, 135–151 (2015).
2. Kaplunov, J., Shestakova, A., Aleynikov, I., Hopkins, B.: Perturbed rigid body motions of viscoelastic structures. In: Proc. IX Int. Conf. Struct. Dyn. EURODYN-2014, Porto, Portugal, 3461–3464 (2014).
3. Milton, G.W., Willis, J.R.: On modifications of Newton's second law and linear continuum elastodynamics. Proc. R. Soc. A: Math. Phys. Eng. Sci. **2079**, 855–880 (2007).

4. Shestakova, A.: Development of mathematical models for freight cars subject to dynamic loading. PhD thesis, Keele University, UK (2015).
5. Kaplunov, J., Manevitch, L. I., Smirnov, V. V.: Vibrations of an elastic cylindrical shell near the lowest cut-off frequency. Proc. R. Soc. A: Math. Phys. Eng. Sci. **472(2189)**, 20150753 (2016).
6. Kaplunov, J., Şahin, O.: Perturbed rigid body motions of an elastic rectangle. ZAMP **71(5)**, 1–15 (2020).

Improving Lithium-Ion Batteries for Electric Vehicles Through Mathematical Modelling

Jamie M. Foster

Abstract The importance of reducing CO_2, NO_x and particulate emissions in mitigating climate change and improving human health is well-known and often stated. Widespread phasing out of vehicles with traditional combustion engines, and replacement with electric vehicles (EVs) is a key step towards achieving a low-carbon economy. Lithium-ion batteries are presently the key technology powering EVs, however, improved performance is highly desirable and would lead to an increase in the rate of EV adoption. Battery development costs can be significantly reduced by robust predictive models, allowing new device designs to be tested *in-silico* without costly and time-consuming physical prototyping. Here, we describe how physics-based battery models have been used to help the industrial sector develop high-performance batteries.

1 Introduction

The lithium-ion battery (LIB) industry is currently worth 44 billion USD per year, and this is predicted to grow to 135 billion USD per year by 2031. LIBs are a key component of electric vehicles (EVs), and are the technology of choice due to their high power and energy density (up to 265 W-hours per kg; around 7 times the density of lead-acid batteries). The LIB comprises $\sim$25% of an EV's weight and accounts for $\sim$40% of its cost, making it arguably the most important component. Consumers are currently hesitant to invest in EVs owing to concerns around battery; (i) lifetime, the warranty period for LIB packs is typically quoted as 100,000 miles or 10 years; (ii) (dis)charge rates, the relatively long time required to "refuel" EVs is tens of minutes rather than a few minutes for combustion engine vehicles; and, (iii) capacity, the limited range per recharge. Increases in battery energy/power density will reduce their weight, meaning longer range. Reducing battery degradation (exacerbated by fast charging) will extend useful lifetime, directly reducing cost to

J. M. Foster (✉)
School of Mathematics & Physics, University of Portsmouth, Portsmouth, Hampshire PO1 2UP, UK
e-mail: jamie.michael.foster@gmail.com

P. J. Aston (ed.), *More UK Success Stories in Industrial Mathematics*, Mathematics in Industry 42, https://doi.org/10.1007/978-3-031-48683-8_12

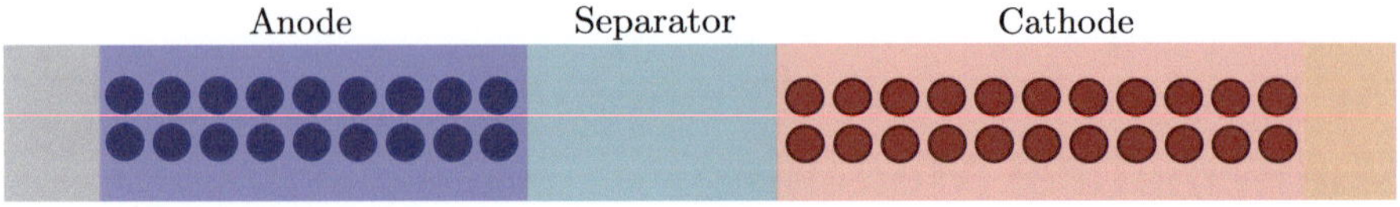

Fig. 1 A single cell, or anode-cathode pair; many of these cells are connected together to form a LIB. The anode, cathode and separator are shown in blue, green and red respectively. The anode and cathode particles are shown in dark blue and dark red respectively. Both the separator and portions of the electrodes that are not occupied by particles are soaked in liquid electrolyte

consumers and reducing environmental harm from disposal and recycling of EV batteries. Such improvements are achieved through better design, and this can be accelerated by supplementing experimental/engineering approaches with robust and predictive simulation tools.

A LIB is made up of a number of connected cells. A single cell, as shown in Fig. 1, consists of a pair of porous electrodes (an anode and a cathode) separated by a porous separator, and bathed in a liquid electrolyte that contains dissolved lithium ions. This assembly is sandwiched between metallic current collectors, used to connect the cell to an external circuit. Each electrode contains microscopic particles of active material into which the lithium ions can intercalate/deintercalate (enter/leave). These particles are connected together by a network of polymer, referred to as binder, which provides mechanical stability and provides pathways for electron conduction. During discharge, lithium ions deintercalate from anode particles and enter the electrolyte. They then flow through the pores in the anode, through the separator, and into the cathode where they travel through the pores and ultimately intercalate into the cathode particles. This transfer of charge leads to a potential difference, which can be used to drive an electronic current through an external circuit. During charge this process happens in reverse; an electronic current is supplied by an external circuit and the lithium ions move from the cathode back into the anode.

Physics-based LIB models are particularly useful as design tools because there is a clear correspondence between model parameters and physical battery components. Thus, once design improvements are identified in simulation they can be translated into practice. The gold standard in physics-based models is the so-called Newman model; the origin of which can be found in [1], and key developments are given in a series of seminal papers from the mid-90 s by Doyle, Fuller and Newman. The model is comprised of a system of coupled partial differential equations to be solved for the lithium-ion concentration in the anode particles, the cathode particles and the electrolyte, as well as the electronic potential in the electrolyte and in the solid network in both the anode and cathode. Much of the authors' impact-generating work is dedicated to: (1) Augmenting the Newman model with the new physics needed to describe LIBs as the materials and designs evolve. (2) Developing software tools that harness modern computing power and allow the fully-fledged Newman model to solved at the scales required to simulate modern pouch cells, roll cells, and even entire battery packs (see www.dandeliion.com [2]). (3) Systematically simplifying, usually via asymptotic methods, the Newman model so that it can be solved at reduced computational cost, and without making significant compromises on accuracy.

2 Mathematical Modelling

A statement of the full Newman model is too lengthy to give here, and so instead we shall now present the so-called single particle model (SPM); the archetypal simplified counterpart of the Newman model. In the form stated here, the SPM can be used to make reliable predictions on the voltage of a cell, V, given the current applied/drawn by an external circuit, I, provided that (dis)charge rates are not too large. An in-depth description of the Newman model, as well as an account of how the SPM can be derived from it, are given in [3].

The lithium ion concentration in the anode (c_a) and cathode (c_c) particles are governed by the following spherically-symmetric diffusion equations

$$\frac{\partial c_i}{\partial t} = \frac{1}{r_i^2}\frac{\partial}{\partial r_i}\left(r_i^2 D_i \frac{\partial c_i}{\partial r_i}\right) \quad \text{for} \quad r_i \in [0, R_i] \quad (i = a, c), \tag{1}$$

where t is time, r_i is the distance from the centre of the particle, R_i is the particle radius, and D_i is the Li-ion diffusivity. The index $i = a, c$, where the former denotes quantities in the anode and the latter in the cathode. Boundary and initial conditions are

$$-D_i \left.\frac{\partial c_i}{\partial r_i}\right|_{r_i=0} = 0, \quad -D_i \left.\frac{\partial c_i}{\partial r_i}\right|_{r_i=R_i} = \frac{j_i}{F}, \quad c_i\big|_{t=0} = c_{i,\text{init}} \quad (i = a, c) \tag{2}$$

where F is the Faraday constant. The former boundary condition is required for spherical symmetry, whilst the latter sets the outward flux of Li on the surface of the particle. The initial concentration in the electrode particles in denoted by $c_{i,\text{init}}$. The quantity j_i is referred to as the intercalation current density and is related to the current supplied by/drawn by the external circuit, I, via

$$j_i = \frac{I}{n_i 4\pi R_i^2} \quad (i = a, c). \tag{3}$$

Here n_i is the number of particles in the electrode. On setting $i = a$ throughout (1)-(3), and having specified D_a, R_a, n_a, $c_{a,\text{init}}$ and I, we have a closed system to be solved for the concentration in the anode particles, c_a, and the current density on their surface, j_a. Analogously, in the cathode, we can determine c_c and j_c having specified D_c, R_c, n_c, $c_{c,\text{init}}$ and I.

Having found c_i and j_i we can determine the surface overpotentials in each electrode (the difference between electrochemical potentials of a Li-ion in the solid electrode material and that of a Li-ion in the electrolyte immediately outside it), η_i, by solving the following equation

$$j_i = 2Fk_i\sqrt{c_i|_{r_i=R_i}\left(c_{i,\max} - c_i|_{r_i=R_i}\right)}\sinh\left(\frac{F\eta_i}{2RT}\right). \tag{4}$$

Here, $c_{i,\max}$, is the maximum concentration of Li in the anode particle, and k_i is a reaction rate constant; both of which are usually determined via experiment. Equation (4) is known as the Butler-Volmer relation and is a staple of surface electrochemistry. Finally, the cell voltage, V, is determined by evaluating

$$V = \left(U_{\text{eq},c}(c_c) + \eta_c\right) - \left(U_{\text{eq},a}(c_a) + \eta_a\right) - IR_e, \tag{5}$$

where $U_{\text{eq},i}(c_i)$ is the equilibrium potential and R_e is the resistance of the electrolyte. Again, both these quantities are usually determined by experiment.

On simulating several battery discharges, each with a different current demand, and using the parameters documented in [4, 5], the SPM predicts the cell voltages shown in Fig. 2. In panel (a) we observe that the higher the current the demand, the shorter the time that the cell can sustain discharge; precisely as expected. However, when we examine how the voltage varies with capacity accessed, see panel (b), we notice that as the discharge rate increases the cell's accessible capacity decreases, i.e., we cannot make use of all of the charge stored within the cell at high rates. Moving to panel (c) we can see that the root cause of this under-utilisation is that the surface of the cathode particle has become saturated with Li-ions (transport from the surface to the interior is too slow). Since the particle surface is full, intercalation cannot continue and discharge ceases. These simulations suggest that the rate limiting process, for this particular cell, is transport of Li-ions in the cathode particles. Therefore, the following design suggestion arises: Rate performance can be increased by manufacturing smaller cathode particles thereby reducing the time needed for Li-ions to travel from the particles surface to its interior, removing the rate-limiting step.

3 Impact and Outlook

By way of conclusion, we now discuss several more examples of instances where mathematical modelling has provided useful insight into how LIB design might be improved, before making some remarks about future work.

As LIBs are repeatedly cycled (charged and discharged) the electrode particles swell/contract causing microscale mechanical stresses which play an important role in battery degradation. However due to LIBs' highly complex morphology, these deformations can be difficult to quantify meaningfully. Motivated by state-of-the-art Scanning Electron Microscope (SEM) images, we developed a systematic homogenisation approach where microscopy images of real battery electrodes with a complicated structure are distilled down to a set of scalar measures (effective transport

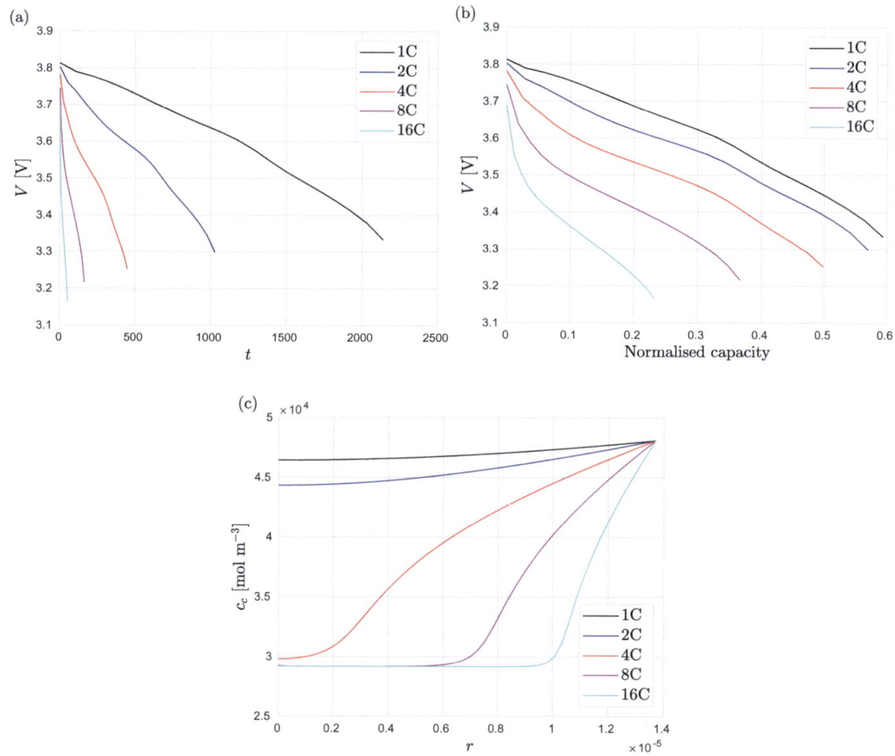

Fig. 2 Results obtained from solving the SPM, (1)–(5), at several discharge rates (C-rates); the higher the C-rate, the higher the current demand. Panel **a** shows cell voltage versus time, **b** shows cell voltage versus (normalised) accessed capacity, and panel **c** shows the concentration within the cathode particles when discharge terminates. Notice that as the C-rate increases, the amount of Li-ions in the cathode particles at the end of discharge decreases

properties) that can be tracked over time; thereby allowing us to more readily quantify changes that occur within batteries as they are cycled, assessing their health and allowing us to suggest when they might be taken out of service. Further, our predictive mechanical models have been used to identify operating protocols and designs that enhance the LIBs' ability to withstand this internal swelling/contraction. In particular, we have given recommendations on how to best alter the shape and material properties of internal components and the manufacturing processes required to mitigate structural fatigue and extend usable lifetimes.

An industrially-relevant strategy for increasing battery capacity is to increase electrode thickness. However this makes Li-ion transport pathways in the electrolyte longer causing electrolytic transport to become the bottleneck and hampering high rate (dis)charge. Together with experimental and industrial partners we modelled bespoke charging protocols that partially overcome this difficulty; a brief current reversal during (dis)charge helps utilise more fully the LIBs' capacity and allows shorter recharge times. A related investigation examined the formation of dendrites, i.e., metallic Li deposited on the surface of the anode particles, typically prevalent at high rate and/or low temperature. When Li dendrites grow to such an extent that they penetrate the separator, they provide a pathway for electron conduction directly between the two electrodes (i.e. a short-circuit), causing thermal runaway and even catastrophic explosion. We developed novel models of dendrite formation in LIBs, giving new insights into how these structures grow over time and paving the way to identify strategies for mitigating them.

Many of our models and solution techniques are embedded within an extremely fast linear-scaling code, namely DandeLiion [2]. Owing to DandeLiion's speed and power, users can perform simulations of batteries and modules that are not possible on other platforms. DandeLiion provides a digital twin of real batteries allowing industry to cheaply and quickly benchmark novel designs, saving significant development time and money. There is ongoing activity to export DandeLiion across academia and industry to help guide the future of LIB development. Readers are invited to test the code (for free and "in the cloud") and follow the progress of the project by visiting www.dandeliion.com.

Finally, we note that despite all this progress there are still significant hurdles to be overcome in terms of (i) improving LIB lifetime, (ii) facilitating higher (dis)charge rates, and (iii) increasing capacity. These developments are needed urgently in order that we realise the multitude of policies that aim to ban the sale of combustion engines across large portions of the world over coming decades (see e.g. the UK's Road to Zero policy). Moreover, EVs are seen as part of the solution for two of the four Grand Challenges in the UK's Industrial Strategy: helping reduce carbon emissions and generating clean growth and changing the future of mobility. Ultimately, better battery design will facilitate vehicle electrification and make a major contribution towards societal goals to reduce harmful emissions, mitigating climate change and reducing the effects of other pollutants on human health.

Acknowledgements The author would like to thank both the Faraday Institution, https://faraday.ac.uk/ EP/S003053/1 [Grant Number FIRG003], and the Engineering and Physical Sciences Research Council [Grant Number EP/T000775/1] for funding parts of this research.

References

1. Newman, J. and Tiedemann, W., 1975. Porous-electrode theory with battery applications. AIChE Journal, 21(1), pp.25–41.
2. Korotkin, I., Sahu, S., O'Kane, S.E., Richardson, G. and Foster, J.M., 2021. DandeLiion v1: An extremely fast solver for the Newman model of lithium-ion battery (dis) charge. Journal of The Electrochemical Society, 168(6), p.060544.
3. Planella, F.B., Ai, W., Boyce, A., Ghosh, A., Korotkin, I., Sahu, S., Sulzer, V., Timms, R., Tranter, T., Zyskin, M. and Cooper, S., 2022. A continuum of physics-based lithium-ion battery models reviewed. Progress in Energy.

4. Ecker, M., Tran, T.K.D., Dechent, P., Käbitz, S., Warnecke, A. and Sauer, D.U., 2015. Parameterization of a physico-chemical model of a lithium-ion battery: I. Determination of parameters. Journal of The Electrochemical Society, 162(9), p.A1836.
5. Ecker, M., Käbitz, S., Laresgoiti, I. and Sauer, D.U., 2015. Parameterization of a physico-chemical model of a lithium-ion battery: II. Model validation. Journal of The Electrochemical Society, 162(9), p.A1849.

Health

Real-Time Epidemiological Modelling During the COVID-19 Emergency in Wales

Mike B. Gravenor, Mark Dawson, Ed Bennett, Ben Thorpe, Carla White, Alma Rahat, Daniel Archambault, Noemi Picco, Gibin Powathil, and Biagio Lucini

Abstract The sudden outbreak of the COVID-19 pandemic presented governments, policy makers and health services with an unprecedented challenge of taking real-time decisions that could keep the disease under control with non-pharmaceutical interventions, while at the same time limit as much as possible severe consequences of a very strict lockdown. Mathematical modelling has proved to be a crucial element for informing those decisions. Here we report on the rapid development and application of the Swansea Model, a mathematical model of disease spread in real time, to inform policy decisions during the COVID-19 pandemic in Wales.

M. B. Gravenor
Swansea University Medical School, Swansea University (Singleton Campus), Singleton Park, Swansea SA2 8PP, UK
e-mail: M.B.Gravenor@Swansea.ac.uk

M. Dawson · E. Bennett · B. Thorpe
Swansea Academy of Advanced Computing, Swansea University (Bay Campus), Fabian Way, Swansea SA1 8EN, UK
e-mail: E.J.Bennett@Swansea.ac.uk

B. Thorpe
e-mail: B.J.Thorpe@Swansea.ac.uk

C. White · A. Rahat · D. Archambault · N. Picco · G. Powathil · B. Lucini (✉)
School of Mathematics and Computer Science, Swansea University (Bay Campus), Fabian Way, Swansea SA1 8EN, UK
e-mail: B.Lucini@Swansea.ac.uk

C. White
e-mail: C.L.White@Swansea.ac.uk

A. Rahat
e-mail: A.A.M.Rahat@Swansea.ac.uk

D. Archambault
e-mail: Daniel.Archambault@newcastle.ac.uk

N. Picco
e-mail: Noemi.Picco@Swansea.ac.uk

G. Powathil
e-mail: G.G.Powathil@Swansea.ac.uk

P. J. Aston (ed.), *More UK Success Stories in Industrial Mathematics*,
Mathematics in Industry 42, https://doi.org/10.1007/978-3-031-48683-8_13

1 Introduction

As one of the four devolved administrations of the UK, the Welsh Government (WG) has responsibility for health policy. The scientific committee for emergencies in Wales is the Technical Advisory Cell (TAC) which worked closely with SAGE, the Scientific Advisory Group for Emergencies at the UK government level. Mathematical modelling played a crucial role in the early stages of the pandemic, identifying the extent of early spread out of Wuhan, estimating transmission rates, case fatality rates and then providing the key evidence for the first UK-wide lockdown to reduce the impact on deaths (before improved therapy and vaccines were available) and to prevent the first wave overwhelming the capacity of the National Health Service (NHS) [1]. The UK lockdown suppressed virus transmission significantly, to the point when restrictions were relaxed during summer 2020, and Wales reported only a handful of cases per day. At that point, there remained great uncertainty on the future trajectory of the epidemic; to what extent it would resurge and what policies could be used to manage the next 12 months? The Swansea Model was originally designed and developed to answer these questions.

Among the Authors, MG is a mathematical epidemiologist, with experience modelling measles, malaria, avian influenza, and mad cow disease, including advising large scale control programmes [2]. MG is a member of TAC and the UK Scientific Pandemic Influenza Modelling Group (SPI-M) at which many of the key UK COVID-19 models were presented and discussed. The results from these models were fed directly into TAC and the WG. Following the largely synchronous initial lockdown, some differences in response emerged across the UK. Although general policy remained similar, timings differed and disease dynamics were often out of synch. Additionally, the demographics of Wales include a particularly elderly population, with large areas facing considerable socio-economic challenges. These issues raised the requirement for bespoke models for Wales which could quickly be tailored to the devolved policy response. In addition to the big picture of the likely course of the epidemic in Wales, and how it might be mitigated, each health board was experiencing their own planning challenges (often at different times) and in need of modelling support for day-to-day expected burden of hospital admissions and bed occupancy.

In this environment, MG was approached by colleagues at WG and Public Health Wales (PHW) to see if a policy model could be rapidly developed and deployed. It was serendipitous that BL had already contacted MG to see if expertise from the Mathematics Department (and part of the Supercomputing Wales project) could help; specifically, a small team of research software engineers (RSEs). The involvement of the RSEs was essential to the success of this project. The time pressure that all scientific advisors were working under was tremendous. Where an academic project of this type might be expected to take place over 1 or 2 years, the urgency of the pandemic could change in a matter of days. The RSEs allowed the team to evaluate and trial existing tools where available (to save time on replicating models) and to rapidly adapt the models for bespoke conditions or scenarios, albeit by working

7 days and deep into the nights. This was the start of the 'Swansea Team' which provided modelling support throughout the pandemic, and has expanded to continue to provide mathematical analysis for COVID-19, influenza, Respiratory Syncytial Virus (RSV) and monkey pox. It has been our philosophy throughout to maintain as close a link as we could between three components, the fundamental epidemiology of infectious disease dynamics, the underpinning mathematical framework that allows quantitative questions to be posed on the epidemiology, and a strong link with end users in the NHS, Public Health and Government so that relevant research questions were always kept to the fore.

2 The Model

In mathematical terms, the "Swansea Model" was based on well-established epidemiological theory, using an 'SEIR' framework, and code published by Davies et al. [3]. Populations are represented in compartments, with the flow between compartments modelled by ordinary differential equations. In a simple example, a population of N individuals has Susceptible (S), Exposed (E), Infectious (I) and Recovered (R) compartments, with the infection processes being 'mass-action' and proportional to the product $S \times I/N$ times the person-to-person contact rate. If i distinct age classes are used, then the basic model is:

$$\frac{dS_i}{dt} = -\beta \frac{S_i}{N} \sum_{j=1}^{n} K_{ij} I_j + \nu R_i \ , \qquad \frac{dE_i}{dt} = \beta \frac{S_i}{N} \sum_{j=1}^{n} K_{ij} I_j - \delta E_i \ ,$$

$$\frac{dI_i}{dt} = \delta E_i - \gamma I_i \ , \qquad \frac{dR_i}{dt} = \gamma I_i - \nu R_i \ ,$$

where δ^{-1} is the incubation period, γ^{-1} is the infectious period, ν^{-1} is the duration of immunity, β is the transmission coefficient and K_{ij} is a matrix of mean contacts per day between each age-group. Our model employed 16 age groups, and was also expanded upon to include: time-dependent contact rates, gamma distributions for the incubation and infectious periods, and additional compartments for those individuals that were: asymptomatic infectious (I_a, which are less infectious), pre-symptomatic infectious (I_p), and symptomatic infectious (I_s). Vaccination was modelled by additional compartments S_v and R_v, which removed Susceptible and Recovered individuals, respectively. This transmission model was linked to a separate simple model in 'post-processing' to keep track of clinical events, defined as delayed functions of the incidence of symptomatic infections. Thus, realistic statistical distributions could be included as delay terms to generate the incidence of clinical cases (C), ward hospitalisations (H_w), intensive care unit (ICU) admissions (H_{ICU}), and deaths (D). By including a duration distribution for each clinical event, the prevalence of cases and occupancy statistics for hospital and intensive care beds were also calculated. Clinical

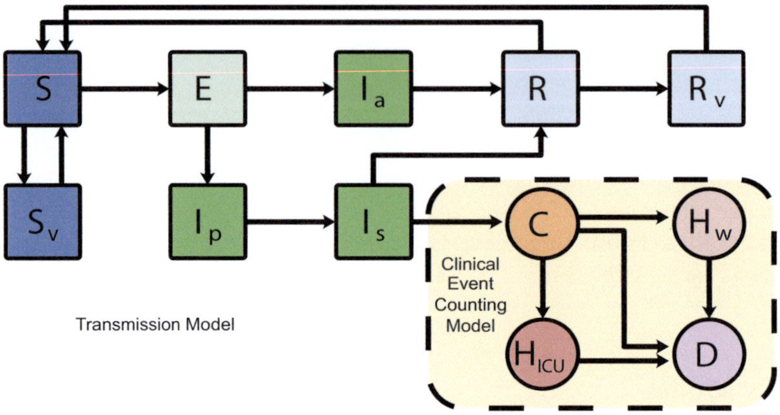

Fig. 1 Schematic of the transmission process, and clinical event counting model

outcomes were essential, since almost all policy target outcomes were events such as cases (for test and trace), deaths, and hospital capacity. Partial vaccine effectiveness was included in the transmission model, as well as different vaccine effectiveness for preventing each of the clinical events. A schematic of the model is shown in Fig. 1. Lastly, we used a separate model for each of the 22 Welsh Local Authority areas, with flexible migration terms to allow for travel, and spread of infected individuals between areas. A stochastic formulation was developed, but rarely used in practice. The model was also compared to a published agent-based model ("Open ABM-COVID 19"), which generated similar results.

Central to the application of the model was a 'schedule' of time- and age-dependent contact rates. Starting with typical contacts obtained from epidemiological surveys, the schedule employed a scaling of these contact rates which represented changes to the typical contacts of the population due to behavioural responses and imposition of non-pharmaceutical interventions (NPIs). Note that the contact rate schedule determines the widely reported effective reproduction number R_t. When R_t was greater than 1 we were facing an exponentially growing epidemic, and for R_t less than 1 the infected incidence was declining.

The contact schedule could be applied retrospectively, by fitting the model to the previous trajectory of the epidemic in Wales, for example from a fit of the model to Public Health Wales time series data on cases, hospital events and deaths. Using Bayesian multi-objective optimisation methods we were able to provide estimates for a large number of parameters including the scaled reduction to 'normal' contact rates, and hence R_t. The next step would be to apply the contact rate schedule prospectively, to simulate the potential future policy scenarios. For example, following a period of strong NPIs such as lockdown, a future scenario could include relaxing of restrictions and an increase in the contact rates. The model could then be used to simulate the future trajectory of key events under those assumptions. Similarly, during periods of alarming epidemic growth, the impact of proposed interventions could be explored by reducing contact rates by an appropriate amount in the schedule. Due to the age-structure, policy interventions could be explored in some detail by teasing apart the effects of movement, work-place, leisure, or school closures; or measures focused on reduction in contacts by the most vulnerable elderly population.

Our mathematical models were used for real time policy advice in several key areas: the estimate of R values; the efficacy of test and trace; assessing the need and impact of winter 2020 NPIs; advising on the likely prevalence of COVID-19 during proposed dates for Senedd elections; informing the road map out of lock-down under the impact of the vaccination roll-out and the emergence of the delta variant; modelling the expected course and impact of the delta wave; estimating the impact of laboratory testing errors on transmission; relaxing test and trace isolation requirements; and modelling emerging variants. Here, we focus on three case studies.

3 The Impact

The Reasonable Worst Case. Summer 2020 was a period of great uncertainty. The first wave had been suppressed and case numbers were extremely low. With few NPIs in place R_t had returned to above 1, but the extent of a resurgence was unclear. We modelled the total population that remained susceptible and factors such as the impact of test and trace, in order to generate a range of winter scenarios. The resulting 'Reasonable Worst Case' (RWC), and 'Most Likely Scenarios' (MLS, based on tracking recent trends and making short-term projections) showed the considerable epidemic potential that remained, how quickly pressures could build and the winter challenges that the NHS would face. An early RWC is shown in Fig. 2. This was published at a time when the risk of any 'second' wave was being questioned and many doubted it could be as large as the first [4]. The sharing of the RWC across services in Wales impacted tactical (3–6 months) and operational (1–42 days) decisions across Welsh hospitals and the ambulance service, providing the basis for staff deployment for: test-trace-protect, in-hospital testing and scheduling, laboratory staffing requirements, and local and government budget allocations. The model scenarios formed part of the central assumptions for hospital bed and critical care capacity planning, elective operating schedules, and procurement decisions in health boards and the NHS Wales Informatics Service, as part of standardised planning assumptions by the third quarter of 2020. Prior warning is required for staffing critical care, and early planning for this reduces overall costs. Based on the model, the need for a 150% increase in capacity was identified and ultimately met when the large second wave was realised.

The Firebreak. By October 2020, the prevalence of COVID-19 was tracking to alarming levels. A widely discussed proposal for delaying and reducing the peak of cases was a short 'firebreak', through NPIs that would reduce the contacts and transmission rates. We modelled multiple different scenarios of possible policy choices coinciding with school half-terms, with different timing, level and duration of the firebreak. Each scenario was evaluated in terms of metrics such as impact on the

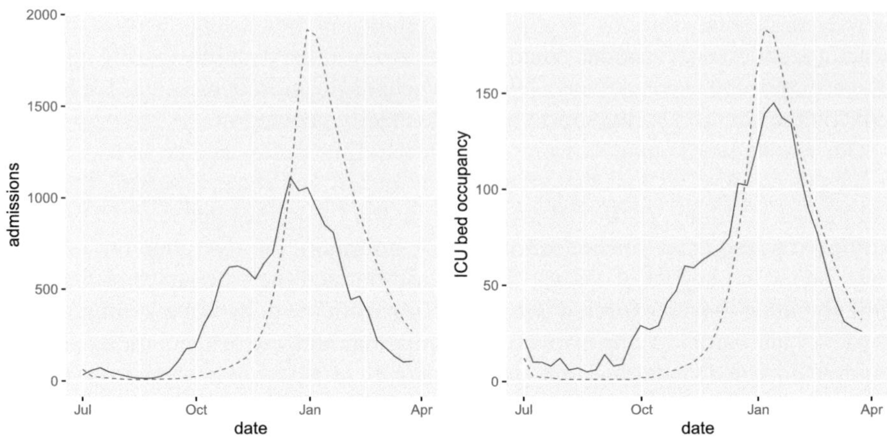

Fig. 2 Winter 20/21 Reasonable Worst Case for weekly hospital admissions (left panel) and intensive care unit occupancy (right panel) in Wales generated in August 2020. Model scenarios are plotted in red (dashed), and the subsequent realised data in blue (solid)

health service, fatalities, pressure on hospital resources and the likely time 'bought', before case numbers would return to the previous levels after NPIs were lifted. The evidence was published simultaneously with the announcement of the 17-day Wales Firebreak on 19th October 2020 [5]. The impact was an effective, if short-term, intervention with an immediate reduction in R_t from approximately 1.4–0.8, resulting in a prevalence 50% lower than England, which Wales had been tracking during the autumn. There was a 6-week period before return to pre-firebreak prevalence, closely following expectations. Vital time was gained to prepare for winter seasonal peaks in hospitals, which would have been harder to manage if cases had continued their trajectory from a higher November baseline. Crucially, cases, deaths, and hospitalisations were significantly reduced (see Fig. 3).

Road-Map Out of Restrictions. Following the second major set of UK NPIs to control the second wave, the UK vaccination programme was initiated. Our efforts then focused on the exceptionally difficult interaction between the impact of the vaccination programme, and the relaxing of NPIs. At this point the models had to be expanded to include new variants, vaccination and waning immunity. A further complication was the differential effects of the vaccine, being highly effective on the more severe clinical events, but less effective at transmission blocking. In early 2021, the models showed that even after the initial dose vaccine roll out to the elderly and vulnerable population, should all NPIs have been removed at that point there remained a very large susceptible population and a high risk of burden of disease and health service impact. These models ultimately informed the staggered road-map out of NPIs, and were used early in 2021 to plan school re-opening as a priority. In January 2021, we modelled staged school re-openings, showing that there would be a small impact on serious event rates as long as other NPIs were prioritised at that time. The youngest

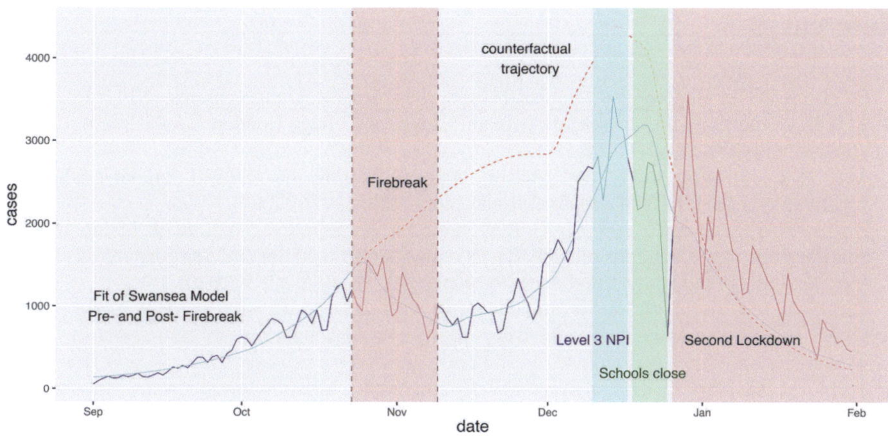

Fig. 3 Counterfactual analysis for Firebreak, contrasting the latter to a scenario without the intervention. Subsequent NPIs were re-introduced to control the December levels of COVID-19

pupils returned to face-to-face teaching on February 22nd 2021 and the rest on March 15th.

Deeper into the vaccine roll out, the model was used to analyse the staged removal of remaining NPIs, and show the limited impact on the most severe clinical events, albeit with a very high expected prevalence of infection due to the size of the remaining susceptible population and the high transmissibility of the circulating virus variants. The combination of 'what-if' scenarios, and short term forecasts based on the model continued to inform the analysis of the delta and omicron variant waves through to 2023, and influence the way winter planning for other seasonal viral infections such as influenza and RSV is conducted.

Acknowledgements The initial year of work was conducted voluntarily and therefore underpinned by our University and Supercomputing Wales posts. Such extensive *pro bono* work highlights an important consideration for future emergency scientific advice planning. We have since had funding to continue and expand the work, and we acknowledge the European Regional Development Fund, EPSRC (grant EP/W01226X/1) and Welsh Government Grants: Sêr Cymru III - Visual Analytics of Epidemic Simulations with Many Actor Behaviours, C098/201/2022, and [SAP]4400015698. We thank Konstans Wells, Miguel Lurgi, Rhodri Fabbro, Finley Gibson and Thomas Torseny-Wier for their contributions. We are indebted to so many colleagues for support on the model and applications: members of TAC, SAGE and SPI-M; especially Brendan Collins, Craiger Solomons, Jenny Morgan, John Watkins, Andrew Nelson, James Cooke, Nick Davies, Rob Orford, Fliss Bennee, Chris Williams, Alun Lloyd, Simon Frost, Rowland Kao, Graham Medley and Laura Andrews. **Research Data and Code Access Statement** Our implementation of the model is available at https://github.com/SwanMod/covid-model. The data generated for this manuscript can be requested from the authors.

References

1. Ferguson, N. et al. Report 9: Impact of non-pharmaceutical interventions (NPIs) to reduce COVID19 mortality and healthcare demand. Imperial College London (16-03-2020), (2020). https://doi.org/10.25561/77482.
2. Kao, R.R., Gravenor, M.B. and Mclean, A.R. Modelling the national scrapie eradication programme in the UK. Mathematical Biosciences, **174**, 61–76 (2001).
3. Davies, N.G., Kucharski, A.J., Eggo, R.M., Gimma, A., Edmunds, W.J., *et al.* Effects of non-pharmaceutical interventions on COVID-19 cases, deaths, and demand for hospital services in the UK: a modelling study. The Lancet Public Health, **5**, e375–e385 (2020).
4. Technical Advisory Cell. New reasonable worst case scenario for winter (18/09/2020), (2020). https://www.gov.wales/technical-advisory-group-new-reasonable-worst-case-scenario-winter.
5. Technical Advisory Cell. Fire Break (19-10-20) (2020). https://www.gov.wales/sites/default/files/publications/2020-10/technical-advisory-group-fire-breaks_2.pdf.

A Risk Prediction Model for Sudden Cardiac Death in Hypertrophic Cardiomyopathy Patients

Rumana Z. Omar and Menelaos Pavlou

Abstract Hypertrophic cardiomyopathy (HCM) is an inherited heart muscle disorder and a leading cause of sudden cardiac death (SCD) in young adults. It affects about 1 in 500 adults in the general population. This chapter describes the first risk model developed to predict the risk of SCD for HCM patients. The risk model has been incorporated into the 2014 European Society of Cardiology (ESC) guidelines and is used worldwide by cardiologists as a risk stratification tool during clinical assessment of HCM patients, to decide on the best treatment option for individual patients. The online risk calculator features on the ESC website. The HCM Risk-SCD model has improved predictive ability compared to the old clinical guidelines offering an optimal treatment decision making tool by balancing risks of SCD and ICD complications. It is a major landmark in the management of cardiomyopathy.

1 Introduction

Hypertrophic cardiomyopathy (HCM) is an inherited heart muscle disorder affecting about 1 in 500 adults in the general population. It is a leading cause of sudden cardiac death (SCD) in young adults and SCD remains the most devastating and feared clinical event for both HCM patients and their cardiologists. HCM patients who are at a high risk of SCD need to be identified so that they can be offered lifesaving treatment with an implantable cardioverter defibrillator (ICD). The old clinical guidelines (2003 American College of Cardiology/ European Society of Cardiology (ESC) and 2011 American College of Cardiology Foundation/AHA clinical guidelines) used to assess the risk of SCD in HCM patients were based on a simple summation of 5 clinical patient characteristics, without the use of statistical modelling. The old guidelines

R. Z. Omar (✉) · M. Pavlou
Department of Statistical Science, University College London, Gower Street, London WC1E 6BT, UK
e-mail: r.omar@ucl.ac.uk

M. Pavlou
e-mail: m.pavlou@ucl.ac.uk

P. J. Aston (ed.), *More UK Success Stories in Industrial Mathematics*,
Mathematics in Industry 42, https://doi.org/10.1007/978-3-031-48683-8_14

had low predictive accuracy (for every 100 patients identified as high risk only 11–23 subsequently had SCD/appropriate ICD shock), resulting in unnecessary and potentially harmful ICD implantation in a substantial number of patients who are at intrinsically low risk of SCD. The objective of this research was to use statistical models to develop a tool that provides more accurate and individualised risk estimates of SCD for HCM patients for a more objective and targeted ICD therapy.

2 Risk Model Development and Validation

Statisticians from the Department of Statistical Science, University College London (UCL) and Cardiologists from The UCL Institute of Cardiovascular Science led on a research study to develop and validate the first risk model to predict the 5 year risk of SCD in HCM patients [1], which is given by

$$\text{Probability of SCD at 5 years} = 1 - 0.998^{\exp\,(\text{prognostic index})}$$

where the prognostic index is the sum of the products of the predictors and their coefficients.

An international cohort study was designed to develop the risk model in collaboration with the HCM-RISK Consortium which includes clinicians and health researchers from Europe, North America, Israel and the Far East. Data on predictors based on patient characteristics and the health outcome, SCD or an equivalent event (aborted SCD during follow-up or appropriate ICD shock therapy) were obtained for 3675 patients from 6 European centres. The survival outcome, time-to-SCD or an equivalent event was modelled using Cox's proportional hazards regression. Patients were followed up for a median of 5.7 years and by the end of the study 5% of the patients had experienced the outcome. The follow-up times of the remaining patients who were either alive at the end of the study or lost to follow up during the study period without experiencing the outcome were regarded as censored in the analysis. First the risk model was developed using the entire cohort and its performance was internally validated using bootstrapping. Subsequently an external validation of the model with independent data was carried out by dividing the cohort into separate model development (using patients from 5 participating centres, n = 2082 with 109 SCD events) and validation cohorts (using the UK Heart Hospital, n = 1593 with 89 SCD events). The calibration slope and the C-index were used to assess the predictive performance of the risk model. The calibration slope assesses the degree of agreement between the observed and predicted hazards of SCD, with a value close to 1 suggesting good calibration. The C-index assesses how well the model discriminates between patients with high and low risks of SCD and a value of 0.5 indicates no discrimination and a value equal to 1 indicates perfect discrimination.

Typically, risk models used in clinical practice are developed without addressing the problems of high levels of censoring often present in survival data, small sample sizes resulting in over-fitting of the model and missing values for the predictor variables. Our research on performance measures for survival outcomes showed that the routinely used Harrell's C index, to assess discrimination in risk prediction studies tended to increase for moderate to high levels of censoring [2]. Therefore, it was likely to give an over-optimistic view of a prediction model's discriminatory ability. Uno's C-index was found to perform better in the presence of moderate censoring and was used to assess the discriminatory ability of the HCM Risk-SCD model. The missing predictor values were imputed using multiple imputation (MI) based on chained equations and a total of 25 imputed data sets were generated. To handle bootstrap validation with MI we adopted a strategy whereby bootstrap validation was performed in each imputed dataset and the model performance measures were estimated after correcting for model over fitting. The estimates of the performance measures were pooled across all 25 imputed data sets using Rubin's statistical rule. Bootstrap validation revealed a calibration slope of 0.91 (95% CI: 0.74, 1.08) and C-index of 0.70 (95% CI: 0.68, 0.72) suggesting good predictive ability of the HCM Risk-SCD model. During external validation the number of events in the model development data was considerably reduced, potentially leading to model over-fitting. An overfitted model may appear to work well in the development data, giving accurate predictions and showing good discriminatory ability, but the apparent good performance may not hold in the validation data. An approach to alleviate this problem involves modifying the estimation method for the regression coefficients by using the penalised maximum likelihood method which shrinks regression coefficients towards zero to improve the resulting model's predictive ability. Our research on the evaluation of penalised methods showed that significant improvements were achieved by taking a penalised modelling approach over standard Cox model for rare outcomes, with the Ridge method performing the best [3]. Thus, the model was fitted in the development cohort using penalised Ridge Cox regression.

3 International External Validation Study

For the HCM Risk-SCD tool to be used widely in clinical practice it was essential to validate its performance in larger independent patient groups to ascertain whether the model predictions hold in diverse populations. A large external validation study was carried out in the second phase of the research using data from a geographically diverse cohort recruited from 14 cardiac centres in the United States, Europe, the Middle East and Asia [4]. The study showed that the risk model performed well in this validation cohort and had a higher discriminatory ability with a C-index = 0.70 (95% CI: 0.68, 0.72) compared to that of the old guidelines C = 0.54 (95% CI: 0.51, 0.56). The model also showed good calibration with a calibration slope = 1.02 (95% CI: 0.93–1.12) and good agreement between observed and predicted risks of SCD at 5 years, particularly for the low risk groups (see Fig. 1).

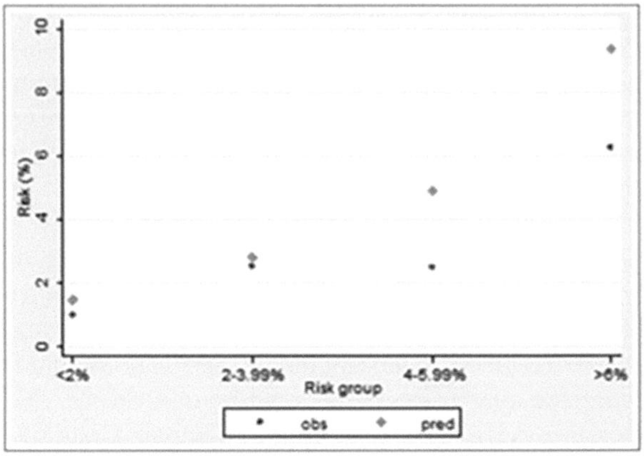

Fig. 1 Calibration by risk group

4 Details of the Impact of the Risk Model

4.1 Impact on Healthcare Policy

The risk prediction tool was incorporated into the 2014 ESC guidelines endorsed in 57 countries [5]. The ESC is a non-profit medical society with member decision-makers drawn from scientists, clinicians, nurses and allied professionals working in cardiology. It unites all national cardiac societies from around the world disseminating evidence-based scientific knowledge to help physicians weigh the benefits and risks of a particular diagnostic or therapeutic procedure. The ESC has classified the risk algorithm as a class I tool (highest of I, II, IIa, IIb, III), which means that the recommendation is based on "evidence and/or general agreement that a given treatment or procedure is beneficial, useful and effective." The risk model's use is recommended to clinicians during the first evaluation and re-evaluation at 1–2 years of HCM patients, as well as whenever there is a change in clinical status. The essential messages released by the ESC on HCM states in their 'take home message': "The use of a new risk calculator (HCM Risk-SCD) is recommended to guide the use of ICDs."

4.2 Impact on Clinical Practice

Published systematic reviews on ICD implantation in HCM patients have estimated that ICD implantation carries a significant risk of major complications requiring intervention or further hospitalisation for 9% of the patients. Furthermore, ICDs can lead to poor quality of life especially when patients receive inappropriate shocks, estimated to occur in a further 11% of patients, which can be painful. Therefore, implantation should only be considered for patients for whom the potential benefits of ICD implant outweigh the potential harms. The old clinical guidelines used for assessing the risk of SCD in HCM patients had low predictive accuracy resulting in unnecessary and potentially harmful prophylactic ICD implantation in a substantial number of patients who are at intrinsically low risk of SCD. The ESC guideline based on the HCM Risk-SCD tool classifies patients into 3 SCD risk groups (low, moderate or severe) using predicted probabilities, thus enabling the identification of the subgroup of patients in whom ICD has the greatest potential to prevent SCD, providing reassurance for the large majority of patients who are at very low risk. The risk tool complements clinical reasoning by providing objective individualised prognostic information and offers improved decision-making for clinicians through more targeted ICD implantation. It is a major landmark in the management of cardiomyopathy.

The Medical Director of the Sarver Heart Center, University of Arizona endorsed the use of the HCM Risk-SCD tool in clinical practice. She stated "The guideline algorithm identifies a group with substantially elevated risk for sudden cardiac death, which is important for patients that are young, as the complications associated with an ICD at a young age are significant." She added "I think the take away message for clinical practice is to become familiar with the [HCM Risk-SCD calculator] for decision making in hypertrophic cardiomyopathy", highlighting that physicians can download an app to their phone or computer.

The tool is used in cardiac centres worldwide and is mentioned in patient referral letters. Clinical leads of cardiomyopathy from UK cardiac centres such as the Oxford University Hospital, the Guy's, St Thomas NHS Foundation Trust and the Northern Ireland Inherited Cardiac Conditions Service stated that the HCM Risk-SCD model is routinely used in their clinical centres to guide risk assessment in their large population of HCM patients and to help rationalise the management of these patients. The ESC survey conducted between February 27th and April 1st 2019, based on responses from 695 cardiologists from 25 countries revealed that there is a widespread use of the tool: 76.3% of the respondents reported using the tool between 1 and 5 times in a 6-month period, 12.5% reported usage of 6–10 times, while the remaining 11.2% reported using it more than 10 times. Approximatively 82% responded that the tool had improved clinical decision making.

4.3 Impact on Patients

It was estimated that for every 13 high-risk patients who receive an ICD as recommended by the ESC guidelines based on this risk prediction tool, 1 patient could potentially be saved from SCD. The HCM Risk-SCD risk tool also saves unnecessary and potentially harmful preventative ICD implantations in patients who are at intrinsically low risk of SCD. The tool provides an informed approach to treatment that empowers patients by helping them understand the relative merits of the primary preventative treatment with ICD and their risk of SCD. For example, patients can weigh their SCD risks against ICD implant decisions, implications of no treatment, and the benefits and risks of treatment with an ICD, and a more personalised care is offered based on their individual risk assessment. The tool thus facilitates patient involvement and a shared decision-making process between patients and their clinicians. This meets the NHS Long Term plan of improving patient satisfaction through the delivery of a better integrated care involving patients, which has been set as a challenge for the government by the Health and Social Care Committee (appointed by the House of Commons). The Chief Executive of Cardiomyopathy UK (the heart muscle charity for patients) stated that "the risk tool is already making a real difference to the lives of people with cardiomyopathy and should be rolled out worldwide so more people can make informed decisions about the treatment options that are best for them." The ESC survey of cardiologists showed that approximately 77% of the respondents routinely discuss the risk estimates from the risk model with their patients and that it helped in their communication with patients.

4.4 Economic Impact

The HCM Risk-SCD tool can be used to avoid unnecessary ICD implants. A very low number of SCDs was observed in HCM patients that had a 5 year predicted risk of SCD of <4% based on the risk prediction tool, supporting the 2014 ESC recommendation not to implant an ICD in these patients [4]. The ICD implantation rate in the UK rose sharply until about 2014 and then has been reducing gradually. England showed a 10% drop in the number of ICDs implanted during 2017/2018 compared to previous years (one ICD device costs the NHS approximatively £10,000, then there is additional cost of hospital equipment and stay) potentially resulting in cost savings. The rate has remained stable from 2018 to 2020 (https://www.hqip.org.uk/wp-content/uploads/2021/10/NACRM-Domain-Report_2021_FINAL.pdf).

4.5 Impact on the Understanding of SCD Risk Factors

The risk model has allowed a better understanding of the relationship between the left ventricular maximal wall thickness (MWT) (one of the risk factors of SCD), and the risk of SCD. Until recently, the clinical perception was that the more severe the hypertrophy (thickening) of the heart muscle, the higher the risk of SCD, and MWT was one of the clinical features used to guide ICD therapy. The risk tool has led to the discovery of a non-linear relationship between MWT and SCD, suggesting that the risk of SCD increases with worsening hypertrophy to reach a plateau and the declines thereafter. This has resulted in an improvement on the guidance of ICD implantations recommending that implantation should not be guided solely on the severity of the left ventricular hypertrophy but a more global SCD risk assessment integrating other risk factors should be used.

Acknowledgements This work was supported by the Medical Research Council grant MR/P015190/1 and the National Institute for Health and Care Research University College London Hospitals Biomedical Research Centre.

References

1. O'Mahony, C., Jichi, F., Pavlou, M.,..., Omar, R.Z., Elliott, P.M.: A novel clinical risk prediction model for sudden cardiac death in hypertrophic cardiomyopathy (HCM-RISK). Eur. Heart J. **35**, 2010–2020 (2013)
2. Rahman, M.S., Ambler, G., Choodari-Oskooei, B., Omar, R.Z.: Review and evaluation of performance measures for survival prediction models in external validation settings. BMC Med Res Methodol. **17**, 60 (2017)
3. Ambler, G., Seaman, S., Omar, R.Z. An evaluation of penalised survival methods for developing prognostic models with rare events. Stat. Med. **31**, 1150–1161 (2012)
4. O'Mahony, C., Jichi, F.,..., Omar, R.Z., Elliott, P.M.: International External Validation Study of the 2014 European Society of Cardiology Guidelines on Sudden Cardiac Death Prevention in Hypertrophic Cardiomyopathy Cardiac Death Prevention in Hypertrophic cardiomyopathy. Circulation **137**, 1015–1023 (2018)
5. Konstantinides, S.V., Torbicki, A.,..., Spyropoulos, A.C.: 2014 ESC Guidelines on the diagnosis and management of acute pulmonary embolism: The Task Force for the Diagnosis and Management of Acute Pulmonary Embolism of the European Society of Cardiology (ESC). Eur. Heart J. **35**, 3033–3080 (2014)

The START (STrAtegies for RelaTives) Intervention for Family Carers of People With Dementia: Evaluation, Statistical Challenges and Impact

Julie A. Barber, Aidan G. O'Keeffe, and Gill Livingston

Abstract The STrAtegies for RelaTives (START) intervention is a psychological therapy to reduce anxiety and depression in family carers of people with dementia living in the community. It was developed and evaluated, using a randomised controlled trial, by a multidisciplinary team including essential contributions from statisticians. This was the first UK trial of a manual-based therapy for dementia carers. The trial randomised 260 carers to START or usual care and assessed them for anxiety and depression over 6 years. Statistical analysis required methods not widely used in applied research to address specific trial complexities, ensuring unbiased and precise effect estimates. Linear mixed models accounted for clustering of outcomes by therapist and joint models for informative truncation of repeated measurements at care-home admission or death. Analyses showed START was clinically and cost-effective in the short- and long-term. The trial impacted UK policy and guidance, being highlighted in the Prime Minister's Challenge on Dementia 2020 and in the National Institute of Clinical Excellence guidelines on Dementia (2018). In 2020 START was being used by at least 24 services across 12 NHS Trusts and by charities including the Alzheimer's Society. It has also been translated and adapted for use in several other countries.

J. A. Barber (✉)
Department of Statistical Science, University College London, Gower St., London WC1E 6BT, UK
e-mail: j.barber@ucl.ac.uk

A. G. O'Keeffe
School of Mathematical Sciences, University of Nottingham, University Park, Nottingham NG7 2RD, UK
e-mail: aidan.o'keeffe@nottingham.ac.uk

G. Livingston
Division of Psychiatry, University College London, 149 Tottenham Court Rd., London W1T 7NF, UK
e-mail: g.livingston@ucl.ac.uk

P. J. Aston (ed.), *More UK Success Stories in Industrial Mathematics*,
Mathematics in Industry 42, https://doi.org/10.1007/978-3-031-48683-8_15

1 Introduction

It is estimated that two-thirds of people with dementia in the UK live at home, typically cared for by family members. This can significantly affect the carer's mental health, with approximately 40% displaying symptoms of clinical depression or anxiety. As the number of people with dementia in the UK is projected to increase to around 1,600,000 by 2040, it is critical to develop strategies to improve the mental health of family carers.

To address this, a multi-disciplinary research team developed and evaluated the STrAtegies for RelaTives (START) intervention; a programme of individualised therapy and coping strategies for dementia carers. In 2009, they conducted a randomised trial of START, assessing its effectiveness at reducing carers' anxiety and depression. 260 carers were recruited and randomly allocated to receive the intervention (known as the 'START' group) or to remain caring for a person with dementia with no additional support (known as the 'treatment as usual' (TAU) group). Carers in the START group received 8 sessions of manual-based therapy individualised to their needs and delivered over 8–14 weeks by 10 therapists (supervised according to specific guidelines for consistency). Outcomes of interest included measures of anxiety and depression, quality of life, health status, carer burden and health care resource use. Follow up data collection was at 4, 8, 12 and 24 months after randomisation and biannually for the subsequent 4 years. The primary outcome was pre-specified as the Hospital Anxiety and Depression Total Score (HADS-T—a validated, numerical score obtained from a series of self-reported questions) over the first 8 months, with the treatment effect of interest being the difference in average HADS-T between the START and TAU groups. This was the first UK randomised trial to consider clinical and cost effectiveness of an individual manual-based therapy for dementia carers. The intervention effect was evaluated in separate analyses focused on short-, medium- and long-term effects [1–4].

Statisticians working on collaborative studies such as START provide essential skills to design an efficient study, address statistical issues arising as the study progresses, and analyse resulting data using optimal, unbiased approaches. Statistical aspects of the START trial involved several challenges requiring application of newly highlighted, non-standard methodologies. Two examples of these are described in the next section.

2 Statistical Challenges

Therapist clustering

In the START group, each therapist delivered the intervention to between 11 and 32 carers. Although therapists were similarly trained, it was considered possible that the intervention effect would be more similar amongst carers treated by the same therapist than amongst carers treated by different therapists. This would result in 'clustering' where outcomes from carers treated by the same therapist are more correlated than those from carers treated by other therapists. Clustering could impact power and precision of the estimated treatment effect and must be accounted for in the study's design and analysis.

Conventional methods for estimating a difference in means in individually randomised trials use simple t-test approaches or analysis of covariance (including adjustment for baseline score (Table 1, Model 1)). These approaches assume that participants' observations are independent. Clustering by therapist would violate this assumption, resulting in under-estimated precision of differences between groups, potentially leading to misleading conclusions about the intervention effect.

At the time of developing the START trial (2007–2014) there was little appreciation in published studies about the potential impact of clustering by health care professionals. For START, however, this issue was raised at the trial's design stage. Anticipating a loss of power, the sample size in the START group was inflated (based on assumed values for the intra-cluster correlation coefficient and average number of carers per therapist), resulting in randomisation of participants to START and TAU groups with ratio 2:1.

The trial's analyses accounted for therapist clustering in the START group while allowing for independence of observations in the control arm. In brief, this involved extending the standard analysis of covariance model (Table 1, Model 1) to a partially nested mixed model including random effects to represent between therapist cluster-

Table 1 Models for analysis of trials with therapist clustering. Equations below are for a continuous outcome y_1, also measured at baseline y_0. i indicates an individual participant, j indicates therapist. t is the treatment group indicator (1 = intervention group, 0 = control), δ is the treatment effect (difference in means), α is the intercept term. Error terms are defined separately for each model

Model 1: ANCOVA model

$$y_{i1} = \alpha + \delta t_i + \beta y_{i0} + \epsilon_i$$
$$\epsilon_i \sim N(0, \sigma^2)$$

ϵ_i = individual participant variation

Model 2: Partially nested homoscedastic model

$$y_{ij1} = \alpha + \delta t_{ij} + \beta y_{ij0} + u_j t_{ij} + \epsilon_{ij}$$
$$u_j \sim N(0, \sigma_u^2),\ \epsilon_{ij} \sim N(0, \sigma_\epsilon^2)$$

ϵ_{ij} = individual participant variation

u_j = a random effect representing the therapist-level variability in the intervention arm

Model 3: Partially nested heteroscedastic model

$$y_{ij1} = \alpha + \delta t_{ij} + \beta y_{ij0} + u_j t_{ij} + \omega_{ij}(1 - t_{ij}) + \epsilon_{ij} t_{ij}$$
$$u_j \sim N(0, \sigma_u^2),\ \epsilon_{ij} \sim N(0, \sigma_\epsilon^2),\ \omega_{ij} \sim N(0, \sigma_\omega^2)$$

u_j = a random effect representing the therapist-level variability in the intervention arm

ϵ_{ij} = individual participant variation in the intervention arm

ω_{ij} = individual participant variation in the control arm

ing in the intervention arm only. Models can be homoscedastic, with the same error variance in the START and TAU groups (Table 1, Model 2) or heteroscedastic, with different error variances in START and TAU groups (Table 1, Model 3). Both models were fitted in analysis of the START data.

Informative censoring

To investigate long-term impacts of the START intervention, carers were followed for 6 years after randomisation with HADS-T recorded at up to 11 time-points. Typically, where there are repeated measurements, long-term treatment effects are estimated using a mixed effects model, including random effects to account for clustering of longitudinal observations within participants. However, this approach may have been inappropriate for the START trial, because 6 years of data were not available for all participants. For many carers, measurements were truncated if their relative died or began to live in full time residential care. Standard models assume that missing data are 'missing at random', where missingness depends only on observed data, but this assumption was unlikely to be valid in this case. HADS-T for carers of people whose condition deteriorates rapidly, leading to death or the need for full-time care, are likely to be different from those who care for people at a less advanced dementia stage who remain living at home for the duration of the trial.

To address this, statisticians proposed a joint model for longitudinal and time-to-event data [5]. The structure of these models is shown in Table 2. For the ith carer, y_{ij} is their HADS-T at follow-up time j and T_i is time until death or care home admission. T_i is a time-to-event (survival) random variable that may be right-censored and the corresponding hazard function is $h_i(t)$.

The joint model accounts for the association between time to care-home admission or death and HADS-T longitudinal observations using a carer-specific random effect, ν_i, shared by two models: one for the time-to-event and the other for HADS-T over time.

The time-to-event sub-model is flexible as the baseline hazard $\alpha_0(t)$ can be modelled using various approaches, including a fully parametric form, spline-based

Table 2 Joint model for HADS-T over time and the time to care home admission or death

Model 4: Joint model
HADS-T outcome:
$$y_{ij} = \mathbf{x}_{ij}^{\top}\boldsymbol{\beta} + \nu_i + \epsilon_{ij}$$
Time-to-event outcome:
$$h_i(t) = \alpha_0(t)\exp\left(\mathbf{z}_i^{\top}(t)\boldsymbol{\gamma} + \phi\nu_i\right)$$ $$\epsilon_{ij} \sim N(0, \sigma_{\epsilon}^2),\ \nu_i \sim N(0, \sigma_{\nu}^2)$$
$\mathbf{x}_{ij}$ = covariates (including the START intervention) affecting HADS-T, with fixed coefficients $\boldsymbol{\beta}$
$\alpha_0(t)$ = baseline hazard function
$\mathbf{z}_i(t)$ = vector of covariates that may influence the time-to-event, with fixed coefficients $\boldsymbol{\gamma}$
ϕ = coefficient of proportionality for the random effect in the time-to-event sub-model

approach or left unspecified. It is also possible to include further random effects, for example, those acting on specific covariates of interest.

3 Results from the START Trial

Results from the START trial were reported in four main papers published in high-impact medical journals between 2014 and 2020 [1–4]. Analyses showed a significant improvement in anxiety and depression for carers eight months after receiving START, with carers in the TAU group four times more likely to have clinically significant depression compared with those in the START group [1]. Corresponding economic calculations demonstrated a high probability of cost-effectiveness [2]. The positive clinical effect of START was sustained after 24 months, with improved carer mood and lower anxiety levels [3]. START remained cost effective with respect to carer and patient outcomes and National Institute for Health and Care Excellence (NICE) thresholds.

Analysis of the 6-year follow-up showed that carers who received START were 5 times less likely to have clinically significant depression than carers in the TAU group over this time [4]. Care costs were nearly 3 times lower for families in the START group (£5,759 per patient) compared with the TAU group (£16,964 per patient), likely due to an improvement in carers' mental health and quality of life [4].

4 Impact

The trial provided robust evidence of clinical and cost-effectiveness of the START intervention and has impacted guidelines, policy and practice for supporting family carers. The intervention manual and resources are freely available from the UCL website which has been viewed approximately 60,000 times up to October 2022.

Changes to UK guidelines and policy on improving dementia care

The UK Government included START in The Prime Minister's Challenge on Dementia 2020 (published in 2015), which set actions to improve dementia care, support and research in England. The government recommended carers of people with dementia should be offered the opportunity for respite, education, training, emotional and psychological support.

The Chief Executive of Alzheimer's Research UK has acknowledged the programme's importance: "It's important to find ways to support carers and protect their health, and these results suggest that the START programme can help reduce anxiety and depression for carers".

START was evaluated in the UK National Institute of Clinical Excellence (NICE) guidelines on Dementia (NG97; 2018), where it was noted that the START trial results provided high quality evidence to support the use of non-pharmacological

interventions, like START, to improve the wellbeing of carers. START has also been mandated by Health Technology Wales guidance (2021) which states "The evidence supports the routine adoption of the Strategies for Relatives (START) intervention for carers of people with dementia".
Delivery within UK

START has generated much interest and enthusiasm from NHS trusts and is being delivered in one third of London memory clinics. Initial dissemination of START, funded by the Alzheimer's Society, led to training of more than 200 clinical psychologists and admiral nurses in several UK locations. These professionals cascaded training to others in their area. A survey in July 2018 showed that START was being used by at least 24 services over 12 NHS trusts, with individuals reporting a better understanding of dementia making it easier for them to cope with psychological distress, improving their wellbeing and ability to provide care.

Worldwide implementation of the START intervention

START has been adopted by the international community, having been translated into Japanese, Chinese, Spanish, Urdu, Tamil, Hindi and Macedonian it is being used in Japan, Hong Kong, Spain, Macedonia, India and South Asian communities. An online version was also developed in Australia to enable delivery in remote areas.

Other impacts

The UCL Institute of Neurology has trialled START for patients with Parkinson's Disease dementia and it has been adapted for people with Lewy Body Dementia. Furthermore, the success of START built a team and a method that led to funding of further related research including the DREAMS-START intervention to improve sleep in people with dementia and a project to improve their quality of life.

References

1. Livingston G., Barber J., Knapp M., Griffin M., King D., Livingston D., Mummery C., Walker Z., Hoe J., Sampson E.L., Cooper C.: Clinical effectiveness of a manual based coping strategy programme (START, STrAtegies for RelaTives) in promoting the mental health of carers of family members with dementia: pragmatic randomised controlled trial. BMJ. **347:f6276** https://doi.org/10.1136/bmj.f6276 (2013)
2. Knapp M., King D., Romeo R., Schehl B., Barber J., Griffin M., Rapaport P., Livingston D., Mummery C., Walker Z., Hoe J., Sampson E.L., Cooper C., Livingston G.: Cost effectiveness of a manual based coping strategy programme in promoting the mental health of family carers of people with dementia (the START (STrAtegies for RelaTives) study): a pragmatic randomised controlled trial. BMJ. **347: f6342** https://doi.org/10.1136/bmj.f6342 (2013)
3. Livingston G., Barber J., Rapaport P., Knapp M., Griffin M., King D., Romeo R., Livingston D., Mummery C., Walker Z., Hoe J., Cooper C.: Long-term clinical and cost-effectiveness of psychological intervention for family carers of people with dementia: a single-blind, randomised, con-

trolled trial. Lancet Psychiatry. **1(7)**, 539–548, https://doi.org/10.1016/S2215-0366(14)00073-X (2014)
4. Livingston G., Manela M., O'Keeffe A., Rapaport P., Cooper C., Knapp M., King D., Romeo R., Walker Z., Hoe J., Mummery C., Barber J.: Clinical effectiveness of START (STrAtegies for RelaTives) psychological intervention for family carers and the effects on cost of care for people with dementia: six year follow-up of a randomised controlled trial. Br. J. Psych. **216(1)**, 35–42, https://doi.org/10.1192/bjp.2019.160 (2020)
5. Rizopoulos D.: Joint Models for Longitudinal and Time-to-Event Data With Applications in R. CRC Press (2012)

New Approach to Modelling of Abdominal Aortic Aneurysms Subjected to Endovascular Sealing

L. P. Argani, R. K. Fisher, R. G. McWilliams, A. B. Movchan, F. Torella, M. L. Wall, and A. Yafawi

Abstract This work is based on mathematical modelling of the failure of endovascular sealing (EVAS) of abdominal aortic aneurysms, with the main aim of an accurate quantitative evaluation of the dynamic response of the EVAS system comprising the biological vascular tissues, fluids and EVAS endoprostheses.

1 Introduction: Outline of the Impact Case Study

The impact case studies discussed here are linked to the medical procedure of the endovascular aneurysm sealing (EVAS). The new modelling approach was developed by the authors in [1]. The code, developed during the study [1, 2], was used in the analysis of the medical data, combined with the results of the medical modelling, which had provided a valuable classification of EVAS structural configurations. During EVAS, the aortic lumen is filled with a bio-compatible polymer, traversed by two 10 mm PTFE-covered cobalt chromium stents. Early clinical results were

L. P. Argani · A. B. Movchan (✉)
Department of Mathematical Sciences, University of Liverpool, Peach Street, Liverpool L69 7ZL, UK
e-mail: abm@liverpool.ac.uk

L. P. Argani
e-mail: luca.argani@unitn.it

L. P. Argani
Department of Civil, Environmental and Mechanical Engineering, University of Trento, via Mesiano 77, 38123 Trento, Italy

R. K. Fisher · R. G. McWilliams · F. Torella · A. Yafawi
Liverpool Vascular & Endovascular Service, and Department of Radiology, Royal Liverpool University Hospital, Prescot Street, Liverpool L7 8XP, UK
e-mail: fcmtdr@liverpool.ac.uk

M. L. Wall
Russells Hall Hospital, Dudley NHS Trust, Dudley DY1 2HQ, UK
e-mail: michaelwall@nhs.net

P. J. Aston (ed.), *More UK Success Stories in Industrial Mathematics*,
Mathematics in Industry 42, https://doi.org/10.1007/978-3-031-48683-8_16

promising [3], but, despite initial assumptions, during follow-up, some patients displayed significant changes in the appearance of the sealed AAA, including migration of the stents/endobags (axial displacement, usually in a caudal direction) and endobags separation [4, 5] (as detailed in Fig. 1). In the papers [1, 2], it was noted that the inertia of the EVAS-treated AAA was a significant factor in possible migration and/or separation of the endoprostheses. The polymer used to fill the endobags surrounding the stent grafts has a relatively high mass density of 2000 kg/m^3 and, as a result, undesired motion of the EVAS components could be observed under dynamic loads associated with activities of daily living. The next section outlines the mathematical model used in the numerical simulations, as well as the main results, which also include the classification of the patient specific EVAS types.

2 The Mathematical Modelling Approach and Main Results

The mathematical approach, developed in [1, 2], includes a static analysis of the EVAS system, subjected to gravity, as well as a formal spectral analysis of the model system (with no dissipation), which provides a vital information about the vibration modes as well as the resonance frequencies. Three case studies have been analysed in [1], which include a simplified model geometry, as in Fig. 1, and two practical cases corresponding to two sets of data for anonymised patients treated with EVAS. In the present paper, a representative selection of ten case studies (illustrated in Fig. 2) is investigated; the geometries and the properties of the EVAS systems associated with the AAA are detailed in Fig. 3.

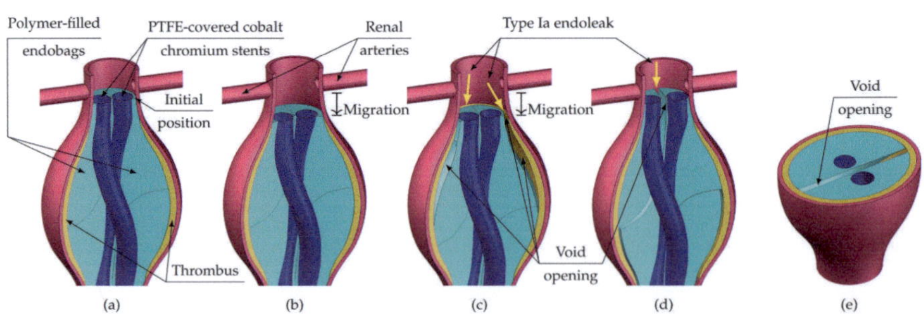

Fig. 1 **a** Schematic representation of the initial position of EVAS components just after installation. **b** Representation of stent migration with no loss of seal and **c** with loss of seal and opening of voids at the top, which allow blood to re-enter the AAA (endoleak). **d** Representation of stent separation with void opening at the top and possible endoleak

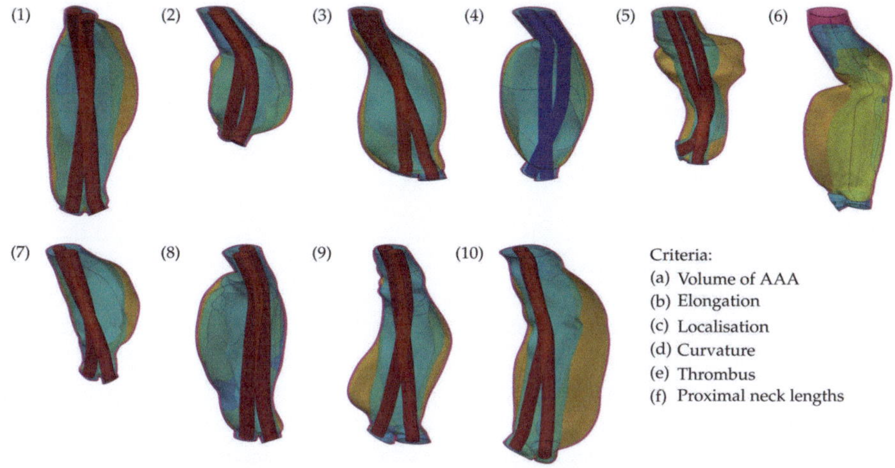

Fig. 2 Geometries of patient-specific AAA treated with EVAS: important parameters, which affect the performance of EVAS include volume of AAA, elongation of the aneurysmal region, localisation of the bulging area, curvature of the exterior and interface boundaries, the amount of thrombus and the proximal neck length

Patient	Maximum neck diameter [mm]				Neck height [mm]		Neck angulation [degrees]	Maximum sac diameter [mm]	Sac volume [mm³]	Sac height [mm]	Polymer volume [mm³]	Thrombus volume [mm³]
	Proximal		Distal		Vertical axis	Centre lumen line						
	Max	Min	Max	Min								
1	27.7	19.0	28.3	19.3	19.0	22.2	31.3	61.0	179713	149.0	164095	57413
2	29.3	24.9	36.3	25.3	31.0	37.0	30.0	60.0	157592	100.0	111642	14283
3	23.2	18.9	25.4	28.0	13.9	15.7	27.8	65.0	193180	124.0	166710	54791
4	30.6	20.4	27.0	23.4	27.9	30.1	22.9	67.0	206310	124.0	185921	33656
5	36.4	40.6	46.7	39.0	31.7	33.0	18.0	81.0	146989	112.2	75159	86039
6	36.0	27.4	42.3	30.6	24.2	28.3	27.0	63.0	188206	145.0	95691	86823
7	33.1	25.7	40.4	31.0	15.9	16.1	12.5	58.7	135589	99.8	87279	26596
8	27.2	21.3	25.8	24.7	21.1	22.3	17.6	63.0	204044	141.5	159020	26555
9	26.4	23.8	33.3	28.3	20.3	22.4	23.0	63.6	170865	142.0	117565	40758
10	37.0	27.3	40.4	36.4	23.7	26.0	23.4	70.0	312409	155.5	127186	189552

Fig. 3 The parameters of the abdominal aortic aneurysms of the selected ten patients at the time of EVAS installation

The governing equations. The displacement field $\mathbf{u}_s$ characterises the motion of the elastic solid components of the AAA repair components, whereas the velocity field $\mathbf{v}_f$ corresponds to the fluid flow, with the fluid-solid interaction being taken into account. Here the notations μ_f, ρ_f and p_f are used for the viscosity, mass density and pressure within the fluid; λ and μ are the Lamé elastic moduli of the solid phase; **f** is the body force density, and σ_s, σ_f are the stress tensors in the solid and fluid, respectively; the vector **n** represents a unit normal on the solid/fluid interface. The linearised model describing the motion of the EVAS treated AAA fluid-solid system is governed by equations $\nabla \cdot \mathbf{v}_f = 0$, $\partial_t \mathbf{v}_f + \rho_f^{-1} \nabla p_f - \mu_f \rho_f^{-1} \nabla^2 \mathbf{v}_f = \mathbf{f}$, and $\mu \nabla^2 \mathbf{u}_s + (\lambda + \mu) \nabla (\nabla \cdot \mathbf{u}_s) = \rho_s \partial_{tt} \mathbf{u}_s$, accompanied by the interface conditions $\mathbf{v}_f = \partial_t \mathbf{u}_s$, $\sigma_s \mathbf{n} = \sigma_f \mathbf{n}$, at the fluid-solid interface boundaries.

Additional considerations for the boundary conditions. At the top and the bottom regions of the sealed AAA structure, the endografts are securely inserted into the artery, which is modelled by spring-type boundary conditions. Frictional contact between the two endobags included in the EVAS system is included in the computational model, as well as frictionless contact between the endobags and the thrombus/aorta. This implies the continuity of the normal elastic displacement across the frictional interface and a constraint on the tangential traction along the interface: for the frictionless contact, the tangential traction is zero, whereas for the case of Coulomb's friction the tangential traction is proportional to the normal traction, with the proportionality factor being the friction coefficient of the value 0.1. Zero tractions are set on the free surface, resulting from the separation of the EVAS components[1].

Classification of patient-specific EVAS types. The shape of an infrarenal AAA may be very different in different patients. Although bulging of the lower part of the aorta is the common feature, elongation of the aorta, which occurs to a variable degree, and the curvature of the affected area are also important in the context of the dynamic response of the EVAS-treated AAA. The following criteria were used in classification of the EVAS types:

(a) volume of AAA;
(b) elongation of the aneurysmal sac;
(c) localisation of bulging;
(d) curvature of the boundary and kinking;
(e) amount of thrombus;
(f) proximal neck length.

Figure 2 includes several types of AAA which will be discussed in the paper. The ten geometries presented in this figure correspond to real patients who have undergone EVAS. These cases are characterised by different overall volumes of AAA, relative elongation in the vertical direction, different locations of the AAA, curvature of the AAA boundary, the length and the diameter of the proximal neck in the upper part of AAA, and the amount of a relatively soft and compliant thrombus contained in AAA. Prediction of the exact dynamic loads acting on the aneurysmal region during a routine life cycle is impossible. However, due to its mass density, the polymer has higher inertia than normal vascular tissues. This brings a new phenomenon of undesirable

[1] Simplified and patient specific models have been programmed into Abaqus Unified FEA (© Dassault Systemes). Hexahedral and tetrahedral shaped finite elements were employed to model solid parts of the sealed AAA. The endografts have a small thickness, and they were modelled by using quadrilateral shell elements.

inertial motion leading to possible migration and separation of EVAS components during activities of daily living. The way to analyse this motion is through the eigenvalue analysis, which also delivers the waveforms and eigenmodes associated with the inertial motion of the EVAS components.[2]

Figure 3 includes the summary of the geometrical parameters of the AAA vascular region for the cases presented in Fig. 2. The following groups of EVAS treated AAA have been identified in the context of their vulnerability to possible migration and/or separation of EVAS components: **(I)** Short proximal neck (Patients 1, 3 and 7). The proximal sealing zone is short, and hence such cases are likely to develop displacement of EVAS components. **(II)** Large AAA volume accompanied by a relatively large amount of thrombus (Patients 6 and 10).[3] **(III)** Significant elongation of the AAA region combined with the regions of high curvature of the exterior boundary (Patients 4, 5, 6, 8, and 9). For such systems, the low frequency modes are associated with the bending/swinging motion, and the regions of high curvature may act as pivoting areas. **(IV)** Relatively small AAA volume, low amount of thrombus and sufficiently long proximal neck (Patient 2).[4]

Clinical observations versus modelling results. Post-EVAS, patients are routinely followed-up with serial imaging, including CT angiography at 1 month and yearly thereafter (with additional scans as dictated by findings). AAA volume and thrombus measurements were obtained from these scans as previously described in [5]. Migration was also measured according to a previously described method of [4], whereas endobag separation was diagnosed qualitatively by visual inspection of the images. We report the results of numerical modelling for the selected patients, compared with the independent clinical observations.

Dynamics of AAA with a short proximal neck. Patients 1, 3 and 7 are characterised by relatively low resistance in the upper region of contact in the neighbourhood of the upper proximal neck. Figure 4 shows the vibration eigenmodes characterised by relatively low first eigenfrequency. In particular, Patient 1 has a swinging lateral motion at 4.35 Hz, and coupled modes accompanied by longitudinal and torsional motions at frequencies of 8.96 and 11.92 Hz. Such low frequencies can be naturally

[2] In the eigenvalue analysis we are interested in the eigenfrequencies in the interval between 1 and 30 Hz, which would be typical for a dynamic input arising from routine activities, such as using common means of public transportation. In the computational model, a constant thickness of 1.5 mm of the arterial wall (for the aorta and common iliac arteries) has been assumed for all the patients. The materials of the polymer components and of the homogenised grafts of the EVAS system have been assumed to be isotropic, linear elastic and homogeneous.

[3] The first eigenfrequency for such vascular systems is low, due to relatively low compliance of the thrombus layer, hence internal displacements of the EVAS treated vascular system are likely.

[4] Such AAA would also have relatively low inertia and hence should not be dramatically affected by transient loads experienced during activities of routine living. On this basis, it is likely to respond well to EVAS treatment and the likelihood of migration and/or separation of endobags is low.

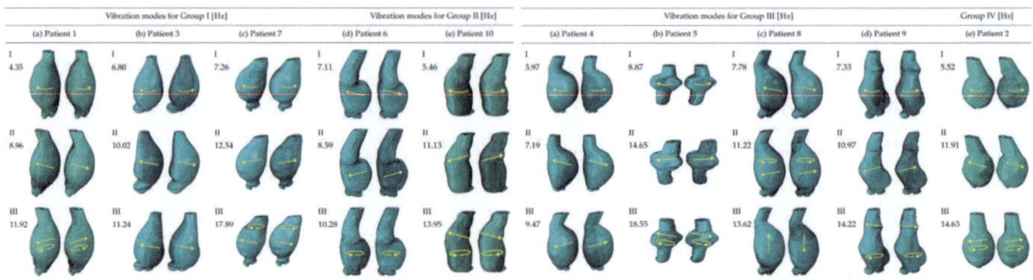

Fig. 4 The first three eigenfrequencies and the corresponding vibration modes for AAA: two groups are characterised by a short proximal neck (group I) and by large volumes with a significant amount of thrombus (group II). Group I: **a** Patient 1, **b** Patient 3, **c** Patient 7. Group II: **d** Patient 6, **e** Patient 10. Two remaining groups are characterised by elongated aneurysmal sacs with the regions of high curvature (group III) and by small volume and sufficiently long proximal neck (group IV). Group III: **a** Patient 4, **b** Patient 5, **c** Patient 8, **d** Patient 9. Group IV: **e** Patient 2

excited in normal life, say while using public transportation or standard physical activities. The mathematical model and simulations suggest a likelihood of migration of EVAS components.[5] Patient 7 has only one eigenfrequency below 12 Hz, but, due to the short, conical proximal neck, the eigenmodes and hence the dynamic response of the entire AAA appears to be prone to lateral swinging at 7.26 and 12.54 Hz and rotational motion at 17.89 Hz. The mathematical model and numerical simulations show longitudinal motion of the endobags, which imply a likelihood of migration of the EVAS components.[6]

Motion of aneurysms of larger volumes with a significant amount of thrombus. Patients 6 and 10 appear in group II associated with the large AAA volumes and large amount of thrombus. These two cases, shown in Fig. 4, have low first eigenfrequencies, with vibration eigenmodes that correspond to swinging lateral motion and rotation. One special feature for both cases is the large amount of thrombus. The mathematical model does not show the long-term evolution of the geometry of the layer of thrombus, hence its prediction applies only for a short time interval.[7]

---

[5] Clinical observations in Patient 1 confirmed that migration occurred after 1 year, with subsequent increase in AAA volume by 29% after 4 years.

[6] Clinical observations showed that the AAA volume increased by 43% at 1 year, with migration being observed as well. Patient 3 represents an intermediate case, where the geometry of the AAA sac and its physical properties lead to a different dynamic response compared to other two cases: in particular, there is no rotational motion for the first three eigenmodes. The prediction of the mathematical model suggests that the likelihood of migration is low, but separation may occur. It was observed that AAA volume increased by 23% at 3 years. There was no migration, in keeping with the theoretical prediction.

[7] Clinical observations show that Patient 6 had migration of the stents components at 1 year, with AAA volume increasing by 65%. For Patient 10 the increase in AAA volume was 16% at 1 year, with subsequent migration of the EVAS components observed at two years.

Significant elongation and high curvature of abdominal aortic aneurysms. This groups includes patients 4, 5, 8 and 9, together with Patient 6. Here, we observe swinging lateral modes, which may occur at low frequencies. Also, rotational modes occur. Patient 4 has the lowest first eigenfrequency due to significant aortic volume and high curvature at the end of the proximal neck. The mathematical model shows longitudinal motion of the stents, which suggests a likelihood of migration. Clinical observations showed migration of the stents at 4.5 years, with AAA volume increasing by 46% after 2 years. Patient 5 has a relatively small AAA, but a high curvature, accompanied by a significant amount of thrombus. The mathematical model did not predict migration and separation of the stents at the upper part of AAA, but the motion of the stents at the lower part of AAA suggested possible separation and a likelihood of endoleak.[8] Although for Patient 8 the mathematical model shows relative displacements of the endobags, clinical observations at 1 year showed minimal migration; the increase in AAA volume at 1 year was 7%. Patient 9 has significant elongation of AAA, and the model has predicted no significant migration or separation of EVAS components. This is consistent with clinical observations, confirming that no migration has occurred; the increase in the AAA volume was 6% after 2 years. For patient 6, both migration and increase in AAA volume occurred.

Aneurysm of a relatively small volume and sufficiently long proximal neck. Patient 2 has a relatively small AAAs. The geometry is shown in Fig. 2(2), and although the curvature of the boundary of the AAA is significant, the length of the proximal neck is sufficient (37 mm) to provide a strong seal in the upper boundary of the AAA.[9]

Summary. The mathematical model has been placed in the context of clinical observations. The computations were performed prior to the clinical comparative study of this group of patients. The results are summarised in Fig. 5, which shows effectiveness of the model. The model does not provide data about the long-term evolution and irreversible inelastic deformations of the thrombus.

[8] Clinical observations after one year showed minimal migration of the EVAS components, but the AAA volume increased by 69% at 2 years; the patient required further intervention.

[9] This case is special, as only two eigenfrequencies are below 12 Hz, and the geometry of the AAA is favourable for EVAS. The amount of thrombus is also relatively small, which reduces the irreversible displacement of the EVAS components. Based on the computations and the mathematical model, there is a limited likelihood of migration due to the vertical displacement of the endografts at higher frequencies above 15 Hz; the likelihood of separation is low. Based on the clinical observations, there was no significant migration (the endograft displacement was less than 4 mm). At 2 years, AAA growth of 22% has been observed.

Patient	Natural frequencies [Hz]						Sac volume [mm³]	Polymer volume [mm³]	Results				
									Computational		Clinical		
	I	II	III	IV	V	VI			Migration	Separation	Migration	Separation	Volume increase
1	4.35	8.96	11.92	14.87	18.98	21.59	179713	164095	Yes	Neutral	Yes	Yes	29%
2	5.52	11.91	14.63	15.70	22.09	24.94	157592	111642	Neutral	No	No	No	22%
3	6.80	10.02	11.24	16.39	20.23	21.09	193180	166710	No	No	No	No	23%
4	3.97	7.19	9.67	11.77	14.69	16.94	206310	185921	Neutral	No	Yes	No	46%
5	8.87	14.65	18.55	23.30	> 30	> 30	146989	75159	No	No	Yes	Yes	69%
6	7.11	8.59	10.28	16.81	19.51	23.46	188206	95691	Neutral	Neutral	Yes	Yes	65%
7	7.26	12.54	17.89	19.37	25.20	> 30	135589	87279	Yes	Neutral	Yes	Yes	43%
8	7.78	11.22	13.62	16.31	20.67	25.09	204044	159020	Yes	Neutral	No	No	7%
9	7.33	10.97	14.22	17.84	22.38	24.47	170865	117565	No	No	No	No	6%
10	5.46	11.13	13.95	15.09	19.18	21.89	312409	127186	Neutral	Neutral	Yes	Yes	16%

Fig. 5 The table provides the summary with the values of the first six eigenfrequencies, the initial volume of the aneurysm, the volume of the polymer deployed in the EVAS procedure, and comparison between the modelling predictions and clinical follow-up observations

Acknowledgements The financial support of the Liverpool EPSRC Centre for New Mathematical Sciences Capabilities for Healthcare Technologies, Grant EP/N014499/1 and EPSRC Impact Acceleration Award through the University of Liverpool (UoL ref. MFG10055) on "Computational modelling for aortic stents" are gratefully acknowledged.

References

1. Argani, L.P., Torella, F., Fisher, R.K., McWilliams, R.G., Wall, M.L., Movchan, A.B. Scientific Reports **7**, 17712 (2017); https://doi.org/10.1038/s41598-017-17759-3.
2. Argani, L.P., Torella, F., Fisher, R.K., McWilliams, R.G., Wall, M.L., Movchan, A.B. Modelling of deformation and dynamic response of abdominal aortic aneurysms treated by endovascular sealing. In: Chakfé, N., Heim, F., Meichelboeck, W. (eds) European Symposium on Vascular Biomaterials (Oct. 12-14, 2017). GEPROVAS Strasbourg (France), 304–305 (2017).
3. Antoniou, G.A., Senior, Y., Iazzolino, L., England, A., McWilliams, R.G., Fisher, R.K., Torella, F. J. Endovasc. Ther. **23**, 285–289 (2016); https://doi.org/10.1177/1526602816628283.
4. England, A., Torella, F., Fisher, R.K., McWilliams, R.G. J. Vasc. Surg. **64**, 306–312 (2016); https://doi.org/10.1016/j.jvs.2016.02.053.
5. Yafawi, A., McWilliams, R.G., Fisher, R.K., England, A., Karouki, M., Torella, F. J. Endovasc. Ther. **26**, 54–61 (2018); https://doi.org/10.1177/1526602818814548.

Healthcare

Transforming Emergency and Urgent Care Services with Mathematical Modelling

Paul Harper, Daniel Gartner, Vince Knight, Geraint Palmer, and Mark Tuson

Abstract Urgent and emergency healthcare requires accurate demand and capacity modelling to ensure the rapid deployment of medical resources and patients quickly assigned to the correct healthcare pathway. A suite of forecasting, queueing, simulation and optimisation models have been developed for capturing time-dependent and time-critical scenarios that have been applied across multiple NHS settings. For example, our research has helped ambulance services to better forecast demand, optimise the geographical allocation of emergency vehicles and roster staff. Our mathematical modelling work was also key to the Welsh Government's investment in the Single Cancer Pathway, which improved nationwide urgent cancer treatment, as well as informing the staff profile required to launch the Welsh NHS 111 service.

1 Introduction

Urgent and emergency medical services play a critical role in providing healthcare. Access to pre-hospital treatment, timely provision of care, and ensuring the appropriate clinical pathway with respect to patient needs can be the difference between life and death. Healthcare pathways typically include stochastic and time-dependent queueing systems with distinct priority classes, for example based on patient urgency.

P. Harper (✉) · D. Gartner · V. Knight · G. Palmer · M. Tuson
School of Mathematics, Cardiff University, Cardiff CF24 4AG, UK
e-mail: harper@cardiff.ac.uk

D. Gartner
e-mail: gartnerd@cardiff.ac.uk

V. Knight
e-mail: knightva@cardiff.ac.uk

G. Palmer
e-mail: palmergi1@cardiff.ac.uk

M. Tuson
e-mail: tusonm@cardiff.ac.uk

P. J. Aston (ed.), *More UK Success Stories in Industrial Mathematics*,
Mathematics in Industry 42, https://doi.org/10.1007/978-3-031-48683-8_17

Our research has focused on modelling healthcare pathways with the aim of addressing critical bottlenecks in care delivery, applying resource where it is most valuable, and identifying opportunities for improvements. In particular, our underpinning research has considered and incorporated three areas as outlined in the following sections.

1.1 Accounting for Variable Priority Levels

Approximation methods are typically utilised to generate staffing requirements for time-dependent dual-class services, limiting the realism and accuracy of modelling. We have developed a numerical approach using mixed discrete-continuous time Markov chains that uniquely account for the variable patient levels of urgency. This approach was embedded in an extension of the Euler method, coined *Euler Pri*, and can incorporate different priority patient classes [1]. Our research has also developed decision models on how to optimally locate and deploy ambulances [2]. Parameters include the most efficient allocation of emergency vehicles to available bases and stand-by points within a region, accounting for predicted demand and patient needs, traffic congestion and routing, and vehicle/paramedic availability. This particular approach is discussed in more detail in Sect. 2.

1.2 Forecasting Demand

Using data from the Welsh Ambulance Service NHS Trust, we determined that forecasting emergency ambulance demand with Singular Spectrum Analysis (SSA) methods was more flexible, and provided superior forecast performance, compared to conventional time series methods [3]. Short-term forecasting of emergency hospital admissions also incorporated weather information using Hierarchical Bayesian models, in collaboration with the Met Office [4].

1.3 Detecting Deadlock

Open restricted queueing networks give rise to deadlock, whereby patients may be unable to leave a server due to upstream blocking, preventing them from moving to the next stages of treatment in the system. This can be very difficult to detect and resolve in a model, while in real life it will often be detected and dealt with by the staff within the system. To improve model accuracy, we have developed *Ciw* [5], an open source simulation Python library that incorporates developed novel deadlock detection by combining Markov models, state digraphs and linear algebraic techniques on transition matrices [6]. This has allowed us to build simulation models more in line with observed behaviours, thus incorporating greater realism.

2 Mathematical Model for Ambulance Allocations with Survival-Maximising Objectives

The geographical locations of emergency medical services (EMS) can directly impact on the probability of patient survival and the overall quality of care provided. The EMS location-allocation problem has been widely studied but has predominantly focussed on minimising response times or maximising coverage and not explicitly for maximising patient outcomes. We have developed a model to consider where to optimally locate different types of EMS vehicles in order to maximise patient outcomes within a heterogeneous population [2]. Below we introduce some aspects of our Maximal Expected Survival Location Model for Heterogeneous Patients.

2.1 Model Formulation

Let m denote the number of demand nodes, n the number of service nodes and k the number of patient types. Using the notation $[a] = \{x|1 \leq x \leq a\}$ for any $a \in \mathbb{Z}$, where $m, n, k \in \mathbb{Z}$, let the demand of type $l \in [k]$ from demand node $i \in [m]$ be denoted by λ_i^l. The average ambulance utilisation at a given service station $j \in [n]$ is given by π_j and the probability that a vehicle at station j is available to respond is therefore $(1 - \pi_j)$.

To maintain an order of preference for all stations in allocation to each demand node, let $\rho \in \mathbb{R}_{\geq 0}^{(m*n)}$ denote the preference matrix; for demand node i, the jth favoured choice of station, ρ_{ij}, having no available vehicle (occurring with probability $\pi_{\rho_{ij}}^{x_j}$, where x_j is the number of vehicles located at the station j) implies the $(\rho_{i,j+1})^{th}$ service node will be the next station to be selected.

It is possible to have a different survival function for each patient type l, $s_l : \mathbb{R}_{\geq 0} \Rightarrow [0, 1]$. For each demand-service node pair, the travel time is required as input to the model, represented by travel matrix $t_{i,j} \in \mathbb{R}_{\geq 0}^{m*n}$. Finally, since the different groups of patients are based on the urgency of the medical condition, a weight w_l for each patient type is required in order to prioritise the demand.

Commencing with the probability of survival of a patient of type l from demand node i, serviced by station ρ_{ij}, the formulation is as follows:

$$P_{i,\rho_{ij}}^l = s_l(t_{i,\rho_{ij}})(1 - \pi_{\rho_{ij}}^{x_{\rho_{ij}}}) \prod_{r=1}^{j-1} \pi_{\rho_{ir}}^{x_{\rho_{ir}}} \tag{1}$$

Therefore, the objective is to maximise the weighted sum over demand nodes for each of the patient groups, $f(z)$ (with specific capacity z), where

$$f(z) = \sum_{l=1}^{k} w_l \sum_{i=1}^{m} \lambda_i^l \sum_{j=1}^{n} P_{i,\rho_{ij}}^l = \sum_{l=1}^{k} w_l \sum_{i=1}^{m} \lambda_i^l \sum_{j=1}^{n} s_l(t_{i,\rho_{ij}})(1 - \pi_{\rho_{ij}}^{x_{\rho_{ij}}}) \prod_{r=1}^{j-1} \pi_{\rho_{ir}}^{x_{\rho_{ir}}} \quad (2)$$

such that

$$\sum_{j=1}^{n} x_j = z \text{ and } x_j \in \mathbb{Z}_{\geq 0} \quad (3)$$

2.2 Survival Calculation

In reality, some emergency incidents do not result in a substantive deterioration of a patient's status over time; however, in all situations, there is a reasonable cut-off beyond which a patient should not expect to have to wait for care. Consequently, a mixture of survival functions and step functions can be utilised by the single allocation optimisation model (2), representing the different objectives (clinical or time focused), as well as the varying levels of patient criticality.

Survival (or *success*) probabilities for critical incidents, calculated from a monotonically decaying survival function representing survival following cardiac arrest [7], are used as an example to demonstrate an attainable level of *success* from a response. This curve $s(t)$ is shown in Fig. 1 alongside a step or hard-target function representing a typical example of an EMS response time target.

When system performance using the hard target approach is translated into survival, the current 8 min UK response target greatly overestimates clinical outcome for

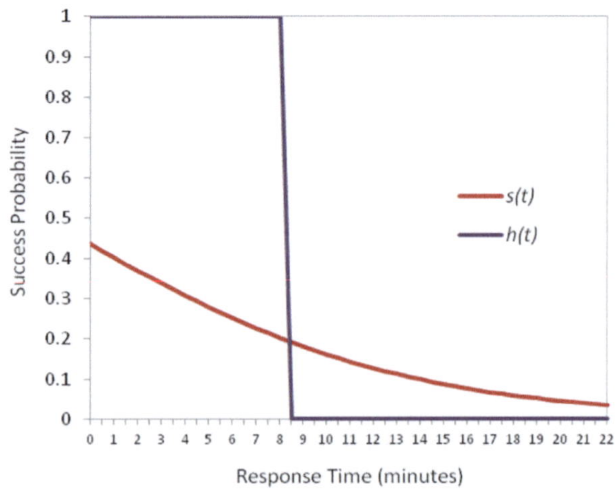

Fig. 1 Illustrative survival function, $s(t)$, compared with a current EMS target, $h(t)$, represented as a step function (binary coverage)

the most critical patients responded to within 8 min, yet predicts no positive outcome whatsoever for patients responded to after this cut-off. The survival curve $s(t)$ offers a more realistic view of the chance of success following a response for more than just cardiac arrest patient groups. Our work has importantly shown that maximising survival as opposed to just minimising response times often leads to different allocations of the EMS fleet and provides enhanced insights for maximising outcomes of the population than using just response time approaches alone [2].

3 Impact

Utilising a range of mathematical modelling approaches, some of which have been described in Sects. 1 and 2, we have worked with a range of NHS partners and Government to help significantly improve healthcare pathway allocation and patient outcomes across NHS urgent and emergency care services in Wales and England.

3.1 Streamlining Pathways to Cancer Diagnosis and Treatment

Working with the Wales Cancer Network and the Welsh Government, our research was used to test the feasibility of a single cancer pathway (SCP). Through the SCP, patient waiting times are measured from when cancer is first suspected (for example, by their GP), not when first referred to hospital, and aims to provide a clear target to speed up diagnosis times and improve survival rates. Using the *Ciw* software [5], we extensively modelled the resources needed to meet the proposed SCP waiting time targets across Wales and quantified the required increase in diagnostic resources to meet proposed treatment timescales, and indicated that rollout of "rapid diagnostic hubs" (where patients with suspected cancer could have a range of cancer screenings undertaken with a single trip) would help achieve the project's goals. The findings were used to advise Vaughan Gething, the Welsh Government Minister for Health and Social Services at that time, who released additional funding to provide the operational and diagnostic support required to transfer to the SCP in line with our recommendations. As a result, Wales was the first UK nation to introduce a single waiting time target for cancer patients, and has been shown to benefit those patients who are now told they do not have cancer more rapidly, and for those who are diagnosed with the disease they can begin effective treatment promptly.

3.2 Supporting Ambulance Services

We have supported both the Welsh Ambulance Service and the London Ambulance Service NHS Trusts, the latter being responsible for responding to urgent and emergency medical situations within the UK's busiest ambulance service. LAS receives 1.9 M 999 calls, and responds to over 1.2 M incidents, annually. Our impact has been across a range of functions, including: (i) development of a forecasting suite that predicts resource requirements as far as 14 days in advance which are used on a daily basis to make decisions around staffing, escalation procedures, anticipated system pressures, and enables informed and proactive decisions at local levels, (ii) increased assurance around demand and performance that has enabled the more effective use of staff time particularly at times of greater pressure, ultimately resulting in a more equitable and safe delivery of service to patients, (iv) has provided services with valuable evidence to confirm the importance of ambulance allocations based on patient outcomes for different conditions and benefits of moving away from target times to survival metrics.

3.3 Designing and Launching the NHS 111 Service

NHS Wales commissioned us to analyse national NHS Direct and Out-of-Hours data to model a proposed 111 service across Wales. The service sought to combine the functionalities of the NHS Direct call-centres with the Out-of-Hours GP phone service and differs from other national 111 services by employing a greater proportion of clinical staff. We evaluated different ways of providing the complex new service and forecasting the optimal staffing size and skills required to support it [8]. Our work quantified the resource needs and allocation, and led to key changes in the operational rollout of the 111 service. Within six months of the initial pilot, the service resulted in 29% fewer ambulance callouts, avoiding unnecessary emergency journeys and enabling Welsh Ambulance to target those in most critical need.

References

1. Vile JL, Gillard JW, Harper PR and Knight VA (2017), A queueing theoretic approach to set staffing levels in time-dependent dual-class service systems. Decision Sciences 48(4): 766–794.
2. Knight VA, Harper PR and Smith L (2012), Ambulance Allocation for Maximal Survival with Heterogeneous Outcome Measures. OMEGA - The International Journal of Management Science, 40: 918–926.
3. Vile J, Gillard J, Harper PR and Knight VA (2012), Predicting Ambulance Demand Using Singular Spectrum Analysis. Journal of the Operational Research Society. 63: 1556–1565.
4. Sahu SK, Baffour B, Harper PR, Minty JH and Sarran C (2014), A Hierarchical Bayesian Model for Improving Short-Term Forecasting of Hospital Demand by Including Meteorological Information. Journal of the Royal Statistical Society, Series A. 177: 39–61.

5. Palmer GI, Knight VA, Harper PR and Hawa AL (2019). Ciw: An open-source discrete event simulation library. Journal of Simulation 13(1): 68–82
6. Palmer GI, Harper PR and Knight VA (2018), "Modelling Deadlock in Open Restricted Queueing Networks". European Journal of Operational Research. 266(2): 609–621.
7. Valenzuela T D, Roe D J, Nichol G, Clark LL, Spaite D W and Hardman RG (2000), Outcomes of rapid defibrillation by security officers after cardiac arrest in casinos. The New England Journal of Medicine 343 (17) 1206–1209.
8. Tuson M, England T et al. (2018), Modelling for the proposed roll-out of the 111 service in Wales: a case study. Health Care Management Science. 21(2): 159–176.

AI Rostering for NHS Doctors in Training: Enhancing Wellbeing with Mathematical Programming

Celia A. Glass

Abstract Rostering is central to the efficient running of any hospital and to the lives of its staff. We show how our rostering program improves staff cover and wellbeing for doctors in training in UK National Health Service (NHS) hospitals. By capturing the mathematical structure of the underlying dynamics, we were able to solve the problem to optimality with Mathematical Programming. The program was run with MPL-Gurobi optimisation software in the Cloud. A pilot study at a London Emergency Department demonstrated reduction in doctors' fatigue, locum staff costs and rota manager's time. Consistent staff cover improved doctors' morale, and both contributed to better patient care. For surgical departments, carefully designed team rosters doubled training time and improved the continuity of care for the patients. These results demonstrate how training, progression and retention in the NHS can be significantly improved. This win-win-win scenario, for hospital, doctors and patients, provides a paradigm shift for rostering in hospitals.

1 Introduction

Good rosters ensure consistent staffing levels and protect staff wellbeing. The existing rostering system for doctors in the NHS (UK's National Health Service) is based upon 'rotas', which are depersonalised rotational rosters, in which flexibility in taking leave comes at the expense of staffing levels. The rotas ensure adequate staff coverage and simultaneously determine whether doctors receive sufficient training and adequate rest. Attempts at protecting the wellbeing of doctors in training, since 2016, have led to increasing complex rules on rota patterns which has made them very difficult to design. Moreover, pressure of work in the hospital has led to an increase in the number of doctors on Less Than Full Time (LTFT) contracts, yet the rota system is designed to handle full time staff only.

C. A. Glass (✉)
Bayes Business School, City, University of London, 106 Bunhill Row, London EC1Y 8TZ, UK
e-mail: c.a.glass@city.ac.uk

P. J. Aston (ed.), *More UK Success Stories in Industrial Mathematics*,
Mathematics in Industry 42, https://doi.org/10.1007/978-3-031-48683-8_18

Retention and progression of doctors in training is vital to any health service. Even amongst highly motivated doctors in the NHS, retention has become an issue due to exhaustion, inflexible work patterns, and insufficient training time. Of particular concern is the slow progression of doctors to consultant grade which is attributed to: tiredness while sitting examinations from lack of sleep; missing specialist workshops due to pressure for ward cover; and insufficient time with specialist consultant in the operating theatre.

We set out to address these rostering challenges through an algorithm for 24/7 working tailored to doctors in training. This required quantifying fatigue and then developing a rostering algorithm to minimise staff fatigue while maintaining staffing levels. Customisation entailed capturing aspirational good working practices and then incorporating them into the rosters. Features include:

- contractual regulations and good practices relating to workload
- personal leave preferences, study leave, training courses and examinations
- LTFT working, team working and cross-rota working.

Rostering 24/7 staff had been the subject of Dr. Knight's Ph.D. research [1, 2]. Dr. Knight unearthed fatigue factors used by the UK Health and Safety Executive (HSE) which we successfully incorporated into a Mathematical Program (MP). The rostering problem is known to be NP-complete. The complexity of the problem thus indicates that it cannot be expected to be solved to optimality and that an heuristic method would be required. However, we found that Mathematical Programming could be applied to gain an optimal solution, despite the problem instances being large. We attribute this to having succinctly captured the underlying structure and specific dynamics, made possible by Dr. Knight's decades of experience in rostering call centres. Even more surprising was the absence of a trade-off with employment costs. This favourable outcome, ironically, can be attributed to the vast size of the solution space.

Expert knowledge was essential for tailoring the rostering tool to cohorts of doctors on 4- and 6 month placements. Dr. Bryan Todd at Horton and Oxford Hospitals and Dr. Duncan Carmichael at Whittington, London, both clarified working practices and reviewed many iterations of rosters until satisfied with the outcome. As well as full coverage of duties for the department, functionality included personalised study and training leave. The program, Nightglass AI rostering, was piloted in Whittington Hospital, London, Emergency Department over 8 rota periods. Subsequent consultancy addressed surgical trainees, reported in REF impact case study [3], and Covid emergency rosters. Several generic features were built into the program, the most important being for LTFT staff and hospital wide rostering.

2 Suitability of Mathematical Programming for Solving Rostering Problems

Rostering is the combinatorial problem of assigning staff to shifts in such a way as to cover the work, lending itself to being expressed in 0–1 decision variables. Controlling the number of staff on a shift is a summation constraint. The legal and contractual employment regulations, such as rules relating to rest periods, can also be formulated as linear constraints. Once quantified, our wellbeing objective and personal leave preferences can be captured as linear expressions to be maximized, or alternatively presented as fatigue and leave request dissatisfaction to be minimised. It is therefore natural for the problem of rostering of doctors in a hospital to be formulated as a minimisation Binary Integer Linear Programme (0–1 ILP). However, this Combinatorial Optimisation problem is well known to be NP-complete (one of Karp's original 21 such problems) and so finding an optimal solution using 0–1 ILP algorithms is expected to be practical only for small instances. Over 50,000 decision variables and 1,000,000 constraints, are needed for just 20 doctors and 14 duties covering a 6 month period, and so one would expect to have to resort to heuristic methods to find any good solution, let alone an optimal one. This was our initial expectation. Nonetheless, we carried out empirical investigation with the 0–1 ILP on the larger problem of 600 staff at large 24/7 Call Centres. It proved to be feasible to find an optimal solution, and even more surprisingly the improvement in quality for the employees did not add to costs for the employer! [1]. The underlying mathematical reasons for this remarkable success lies in the nature of the rostering problem formulation whose large sparse 0–1 decision matrix is unimodular and thus amenable to fast solvers.

Integer Linear Programming is well covered in textbooks, and ever evolving solution methods are incorporated within commercial software optimisation packages, such as Gurobi Optimizer, FICO Xpress and IBM ILOG CPLEX. Briefly, Binary Integer Programming is solved with Branch and Bound or Cut technique in which the solution space is recursively divided up, as represented by a binary tree with a branch for each 0–1 variable option. The tree is first scanned from root to leaves in some systematic order. At each node the corresponding Linear Programme (LP), in which the integer variables are relaxed to allow non-integer values, is solved quickly to give a Lower Bound for the objective function value of this whole branch of the tree (LB). The branch is eliminated (pruned) if the LP is infeasible or has a higher value than the best integer solution found so far (the global UB as described below). Each leaf node of the tree, if feasible, represents a possible solution in which all of the 0–1 variables are determined, and thus provides a candidate solution value. The program maintains the currently best solution value as a global minimum Upper Bound to the optimal value being sought (UB). Backtracking from leaves to root of the tree allows further pruning.

A Linear Programming solution space is represented by a multi-dimensional polytope whose facets represent the constraints. The optimal value is known to occur at a vertex point of the polytope, which greatly reduces the search space for a solution. There are therefore fast LP algorithms, from Simplex to Interior-point Methods, which commercial implementations enhance through parallelisation and sophisticated pre-processing.

A major drawback of ILP is that runtimes can become prohibitive, and so ways of speeding up the process are essential. This generally involves decomposition of some sort. One approach is to seed subsequent runs with previously identified good solutions, in the form of column generation. Another is to decompose the problem into two stages, though success has been shown to be very problem dependent, and not appropriate in all rostering contexts. In our case, decomposition was required, and the levels of complexity of the problem was used as it proved to be more appropriate than consecutive time periods.

3 Modelling of the Rostering Problem as an Integer Linear Program

The manual rota system had outlived its usefulness. However, doctors were used to relying upon the rotational nature of rotas to ensure duty coverage and contractual compliance and for perceived fairness. Our approach to achieving these ends was to capture the underlying purpose and broader operational context, instead of replicating the existing rota system.

There are three aspects to medical rotas: legal and contractual regulations; duty cover requirements; and quality considerations. Each required both constraints and objectives. In MP there is not a clear a distinction between constraints and objectives, and their interchangeability is part of the modelling process. Even the legal and contractual constraints can safely and efficiently be handled as terms in the objective function with high weights. The associated individual weights, Lagrange multipliers, provided the Goal Programming feature whereby each objective is optimised individually in turn. In our situation, where the solution space is vast, we were able to achieve the best, or near best, quality value for all objectives simultaneously.

Mark-up languages, such as AMPL, GAMs and MPL, provide an invaluable mechanism for high-level algebraic formulation of an ILP. We used Maximal Software's MPL which unpacked the formulation into individual lower-level ILP constraints for input into the Gurobi Optimizer package. Gurobi then pre-processed the formulation to simplify the instance before applying very powerful solution algorithms. Running remotely in the Cloud facilitated parallel computation. The program was reconfiguration through input variables without reprogramming, and quality parameters were refined within runtime limitations through multiple runs. Further AI (Artificial Intelligence) features ensured that the program automatically adapted to departmental context, intensity of duty requirements, and multiple wards and/or grades.

For doctor rostering, a small instance entered as 93 high level constraints in MPL code, unpacked to over 600 times as many, 62,310, individual constraints (rows), and 45,477 variable (columns). Gurobi reduces the size of the instance with a pre-solver, in this case by about two thirds to 23,020 rows and 16,040 columns using 14,534 binary decision variables. The rostering decision problem was characterised by very sparse MP matrices with data density less than 0.015%.

The following extract of the MPL code, of the quality component of the objective function, illustrates its readability and succinctness.

```
MIN
.........
9 DutyBwdObj               + 6 HrsWorkedPerWeekMaxObj  + !Fatigue
8 LeaveDissatisfactionObj + 7 StudyDissatisfactionObj  + !Leave Preferences
3 FairnessDiffObj                                        !Equitability
```

The first term, DutyBkwObj, captures the good working practice of ensuring that no one is asked to start work earlier on any day than they did the day before, i.e. avoiding 'backward rotation'. It applies to each doctor on each consecutive pair of days in the roster. The workload objective, HrsWorkedPerWeekMaxObj, applies to each doctor in each 7 consecutive days or duties of the roster, to ensure that total working time does not exceed 72 h. Annual leave satisfaction applies to all staff and involves doctors' priority weights on their leave requests. To fit in with the minimisation direction of the objective function, we consider negative of satisfaction, termed dissatisfaction. Allowing doctors to put forward multiple requests for leave weeks with their own priorities worked well. It provided the program with sufficient flexibility to balance doctors' leave requests against each other so that invariably all doctors were assigned leave weeks of their highest priority.

The ILP program structure was designed to facilitate extensions of its basic functionality. For example, the previously unmet need of LTFT contracts could be accommodated by using proportional working hours and restricted working time and days. Similarly, providing cross-cover involved running multiple rosters simultaneously with joint coverage constraints, using duty and doctor attributes relating to ward and grade.

The process took between seconds and hours to run depending upon the complexity of the shift patterns and level of detail of the personal requests. Runtime is not critical for this application as roster schedules are published every 6 months. Day to day updates are handled separately.

Our role went beyond the MP algorithm. To be effective we needed, on occasion, to challenge given interpretation and data. For example, the limit of 'a maximum workload of 72 h in any one week', introduced in 2016, was initially implemented by other software providers as starting at 00:00 on Mondays. It took complaints by doctors of 80 h workloads over 7 consecutive daily shifts for the matter to be resolved. Then managers were unable to adapt their existing rotas to include the weekend prior in an annual leave week, nor a new, apparently innocuous, contractual requirement, for one of the weekends' rest period of 59 h to be extended to 62 h once in every 4 weeks. We accommodated both quality improvements through targeted changes in a few duty times using the MP's feature for identifying a minimum set of conflicting constraints.

4 Impact

We undertook a pilot study in Whittington Hospital, London, Emergency Department, for junior doctors' 4- and 6- month placements, between 2017 and 2019. The benefits were reported by Dr. Carmichael [4]:

> "How Whittington Hospital's new rostering system reduces fatigue for staff, protects staff and patients, and reduces locum spend through redeployment of existing staff. It effectively builds training and personal commitments, while maintaining full service cover."

Maintaining staffing levels through rostering, reduced locum costs and saved in staff time managing the rota, resulting in both a financial saving in excess of £25,000 per p.a. and improved quality of care.

Doctors reported significant improvement in morale from gaining flexibility in leave and special weekends off. This is in the context of previous uncertainty around even a doctor's own wedding day off due to the demands and complexity of rotas. We introduced a fortnight as a standard option for personalised annual leave. The difference for the doctors is highlighted by one who had not seen her family in India for 9 years, as her previous rotas had restricted annual leave to 1 week at a time. She received 3 weeks away with time-zone adaption the following week within her roster, at no inconvenience to fellow doctors.

Autonomy over working times alleviated the pressure on doctors of their intensive work. An amateur musician was able to attend regular performances; a doctor who wished to run a marathon on a Sunday was granted the weekend off-duty and the following Monday morning off as recovery time. The doctors were also grateful for assurance of having teaching course and interview days off duty avoiding the common problem of being placed on conflicting night duty. In particular, being rostered off duty on the 2 nights before an examination proved to be important for career progression. One of the biggest frustrations voiced by doctors, affecting female doctors disproportionately, was the difficulty of working part-time, LTFT. Treating doctors individually, as we do, provided the required flexibility, as well as more efficient staff coverage for the department.

Obstacles to training were particularly apparent for specialist registrars. We looked into this for Kings College Hospital Emergency Department where 9 of the 15 registrars were specialising in Emergency Medicine. Training is organised at a regional level, with monthly day workshops. The original rota assigned the 9 specialist doctors 13 night duties which clashed with the 5 workshop days. Nightglass customised AI roster, by contrast, avoided the conflict, without any compromise in staff cover or deterioration in the rosters of the non-specialist registrars. The cost of locum doctor cover to achieve a similar end was estimates to be over £40,000 p.a. if it could be found. In practice, sufficient suitable locums could not be found as all Emergency Departments in the region were all similarly stretched for the same reason.

Surgical departments were having similar problems. The number of surgeons completing their training had slowed down over recent years. In 2019, I and an anaesthetist, Dr. Azra Zyada, were asked to investigate by the Royal College of Surgeons sponsored by Health Education England [3]. The key factor was the proportion of time working under specialist supervision, which had dropped to below 45% on the rota in a typical London teaching hospital, and needed to be above 60% for adequate training. Through individualised rosters, we were able to pair the trainee doctor with their mentor consultant over 70% of the time. In addition, highly prized operating theatre and clinic sessions were increased by a third, and trainees saw over twice as many patients through from admission to discharge. The roster thus provided an unanticipated benefit of continuity of care for patients, in addition to the significant increase in training.

In 2020 we were brought in to Guys and St Thomas' Hospital to advise on the anaesthetists' rotas during the Covid emergency. Anaesthetists are indispensable for Intensive Care and Birthing wards. However, availability of doctors was unpredictable and reduced, and jobs took longer to perform with Covid prevention procedures. In consequence, all non-sick leave and training were cancelled, as were most surgical operations. As a consequence, anaesthetists were not graduating to consultant level. We addressed the problem by designing rosters for transition out of the Covid emergency which ensured that the senior registrars were on duty for their qualifying specialist operating theatre sessions, once these recommenced in the hospital.

This chapter provides a taste of the complexities of rostering in the NHS, and the power of optimisation to provide excellent solutions. It is gratifying that staff wellbeing, costs and staff coverage all improved, as did progression to consultant. We thus established benefits all round, for doctors, hospitals and patients.

Acknowledgements I wish to acknowledge Dr. Duncan Carmichael of Whittington Hospital, London,, Dr. Bryan Todd, of Emergency Department, Horton Hospital, Oxford Education Deanery, and Judge Andrew Worthley, when a lecturer at City University London. I am particularly indebted to Dr. Azra Zyada, Anaesthetist, University College Hospitals, London for her dedication to the cause of junior doctors' wellbeing. I acknowledge Health Education England, for funding the consultancy contract with Royal College of Surgeons, *Improving Surgical Training (IST) Pilot, March 2018.*

References

1. Knight, R.A.: Optimisation methods for staff scheduling and rostering: An employee-friendly approach. PhD thesis, City, University of London, supervised by Professor Celia Glass. EThOS ID: uk.bl.ethos.516843 (2008)
2. Glass, C.A., Knight, R.A.: The nurse rostering problem: A critical appraisal of the problem structure. Eur. J. Op. Res. **202** 379–389 (2010)
3. City, University of London, REF2021 Impact Case Study: A solution to staff rostering problems in hospitals by means of a 'fair' algorithm. https://results2021.ref.ac.uk/impact (search keyword 'rostering')
4. Whittington report - NHSi(NHS improvement) HEE(Health Education England) 2018 Case study: fatigue-minimising, flexible e-rostering in the Emergency Department. https://city.figshare.com/articles/online_resource/Fatigue_Minimising_Flexible_e-rostering_in_the_Emergency_Department_an_HEE_Case_Study/22241722

Meta-Analysis and Survival Analysis in Healthcare: Methods, Applications and Impact

Yinghui Wei

Abstract We described our novel research in medical statistics and their impact in healthcare by drawing upon the following examples: a meta-analysis within the Cochrane review on the effect of cutting portion size on consumption of food; a Stata command to regenerate individual level time-to-event data from published survival curves; and the use of the difference in restricted mean survival time as an alternative to hazard ratios when analysing time-to-event outcomes.

1 Introduction

Meta-analysis is a statistical method to combine results from multiple studies, which answer the similar research questions. Aggregate data meta-analysis refers to the case when data are available at the study level but not at the participant level. Individual participant data (IPD) meta-analysis refers to the case when individual, raw data are available for the analysis.

IPD meta-analysis is the gold standard of systematic reviews, as it allows to, where appropriate, harmonize the statistical analysis of participant-level data across studies. However, IPD are not always available and can be time consuming to collect, and so has inevitable funding implications.

Time-to-event data provide information on whether an event of interest occurs, and time to the earliest of such an event and censoring. Hazard ratios are often used to measure the effect of an independent variable on a time-to-event outcome. However, the proportional hazards assumption is not always realistic.

We described our methods and applications in meta-analysis and survival analysis, and the impact developed from our research in these areas.

Y. Wei (✉)
Centre for Mathematical Sciences, School of Engineering, Computing and Mathematics, University of Plymouth, Plymouth PL4 8AA, UK
e-mail: yinghui.wei@plymouth.ac.uk

P. J. Aston (ed.), *More UK Success Stories in Industrial Mathematics*,
Mathematics in Industry 42, https://doi.org/10.1007/978-3-031-48683-8_19

2 Meta-Analysis

2.1 Methods for Meta-Analysis

Suppose that there are n independent studies included in the analysis. Denote θ the overall effect of an independent variable on the common outcome of interest across studies. The independent variable may be a treatment indicator in clinical trials or exposure indicator in epidemiological studies, where θ refers to treatment effect or exposure effect, respectively.

In a random-effect meta-analysis model, the observed effect in a study is given by

$$\hat{\theta}_i = \theta + \xi_i + \varepsilon_i, \tag{1}$$

where ε_i is the difference between the common true effect θ and $\hat{\theta}_i$, which is the observed effect estimated from study $i (i = 1, \ldots, n)$. It is assumed that $\varepsilon_i \sim \mathbf{N}(0, s_i^2)$, where s_i^2 is the within-study error variance. Here $\xi_i \sim \mathbf{N}(0, \tau^2)$ is the random effect of study i, where τ^2 refers to the between-study variance. Equation (1) becomes a fixed-effect meta-analysis model when $\xi_i = 0$. The overall effect θ can be estimated by using maximum likelihood, and the between-study variance τ^2 can be estimated by using DerSimonian and Laird method or the method of moments. The meta-analysis model can be extended to analyse multiple outcomes simultaneously, namely multivariate meta-analysis [1].

2.2 Applications of Meta-Analysis

The Cochrane sizing review [2] suggested that cutting portion, packaging and tableware size presented a potential path to tackle obesity, which impacts a quarter of British adults, costing lives and quality of life. Wei co-led the statistical analysis in this review to evaluate the effect of portion, package, individual unit and tableware size on changing the consumption of food, alcohol and tobacco [2]. We conducted a meta-analysis of 86 independent comparisons from 58 studies (6603 participants), which concluded an effect of portion, package, individual unit or table size on consumption of food (standardised mean difference (SMD) 0.38, 95% CI: 0.29 to 0.46), providing evidence that exposure to larger sizes increases quantities of food consumption. Subgroup meta-analyses show that exposure to larger sizes increased quantities of food consumed among children (SMD 0.21, 95% CI: 0.10 to 0.31) and adults (SMD 0.46, 95% CI: 0.40 to 0.52). The size of this effect suggests that, if sustained reductions in exposure to larger-sized food portions, packages and tableware could be achieved across the whole diet, this could reduce average daily energy consumed from food by between 144 and 228 kcal (8.5 to 13.5% from a baseline of 1689 kcal) among UK children and adults.

2.3 *Impact on Government Policy Change in Food and Drink Sectors*

Obesity has become one of the major challenges of our time. It is a complex condition, with significant contributions from genetic, physiological, environmental, and social and psychological factors that influence diet and activity. About two thirds of UK adults are now overweight and nearly one third are obese. Obesity alone accounts for at least 10% of NHS spend, and this is rising. Therefore, the prevention and treatment of obesity has become a national priority. In the UK, NICE guidance "Obesity: Identification assessment, management" identified the broad principles of dietary (and other) treatments, including the aim of achieving an energy deficit while maintaining a healthy and balanced diet. However, the guidance did not provide detailed recommendations on how this can be achieved.

Our Cochrane review and meta-analysis [2] rectified this problem by providing important evidence demonstrating that reducing the size of food portions and tableware can be used as a strategy for weight loss. The review was cited in Public Health England's (PHE) October 2015 publication,"Sugar reduction: The evidence for action report", which had been commissioned by the Department of Health to feed into the Government's childhood obesity strategy. PHE's report highlighted our review to argue for the benefits of reducing portion sizes. This was then used to create a programme of gradual sugar reduction in food and drink combined with the reduction in portion size and led to the development of various tools such as the Eat Well Guide, that assists patients in achieving weight loss. "[The Cochrane sizing review's] evidence-based strategy has become a key principle of weight management and is now firmly embedded within weight management programmes such as ours, and elsewhere in the UK and worldwide". A Consultant Physician from Plymouth and West Devon Weight Management Service, Livewell South West.

Our Cochrane review [2] raised public awareness on portion size and regularly receives media attention in news and television programmes. This has resulted in the concept of portion control being universally recognised as a fundamental requirement for the prevention of weight gain and its associated health comorbidities, and for achieving weight loss.

Internationally, the findings of the review were shared with Australia's Victorian Health Promotion Agency and used in their subsequent citizen jury, which resulted in a recommendation that beverage sizes should be regulated and that a maximum size of soft drinks and other calorie-dense beverages that can be sold through restaurants and retail outlets should be imposed.

3 Regenerating Individual Participant Time-to-Event Data

3.1 *Stata Command* ipdfc

We developed a Stata command `ipdfc` [3] to enable the accurate regeneration of time-to-event data using information extracted from published Kaplan-Meier curves. In such curves, the x-values usually represent the follow-up time; the y-values may represent the survival probabilities at the corresponding time points. Alternatively, the y-values may be survival percentages, failure probabilities or percentages, as specified in the study publications. All these measures can be transformed arithmetically into survival probabilities.

For each curve, we extract survival probabilities S_k at time points t_k, $k = 1, 2, \ldots, K$, where K is the total number of extracted data points; and number of patients at risk $nrisk_i$ at time $trisk_i$, $i = 1, \ldots, T$, where T is the number of intervals where the number of patients at risk is reported. If the risk table is not reported but only the total number of patients in each group (for example, treatment group in clinical trials) is known, we have $T = 1$.

The command `ipdfc` makes use of the extracted data, to first calculate the number of censored indviduals, and the number of events within each extracted time interval $[t_k, t_{t+1}]$, and then distribute individual event times and censoring times evenly within the respective extracted time interval. The accuracy of the regenerated data is further ensured by putting stopping constraints in the algorithm, by requiring the estimated number of people at risk equaled to the observed number at risk, and the estimated number of events not lower than the observed number of events if reported.

3.2 *Application*

Figure 1 shows the reconstructed Kaplan-Meier curves based on the extracted data of EUROPA trial [4], a two-arm randomized placebo-controlled trial evaluating the efficacy of perindopril in reduction of cardiovascular events among patients with stable coronary artery disease. The reconstructed curves correctly reflect that the benefit of perindopril treatment began to appear at one year and gradually increased throughout the follow-up of the trial.

3.3 *Impact on Saving Time and Costs in Data Collection*

The collection of individual raw data is time intensive (often taking months to years), usually requires dedicated staff with a variety of skills and it costs more for an IPD meta-analysis than a meta-analysis of published data. Our innovative statistical package `ipdfc` [3] used published curves to rapidly regenerate individual participant data at little cost.

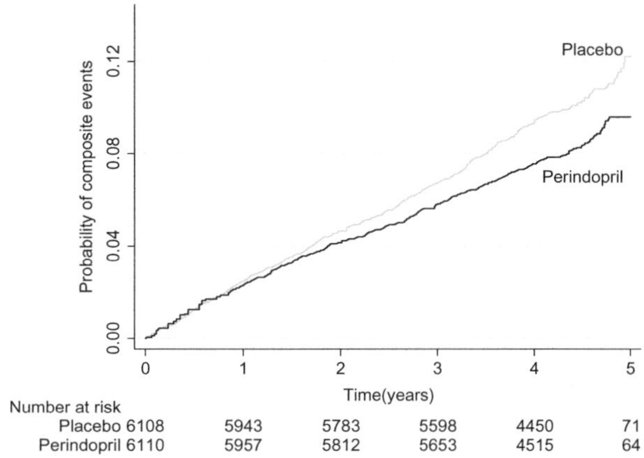

Fig. 1 Reconstructed Kaplan-Meier curves for time to first occurrence of event. Patients are randomly assigned to perindopril treatment or placebo, EUROPA trial [4]

As of November 2020, the package has been downloaded more than 153 times from the Stata Journal site, with users in the UK, USA, China, Spain, France, Netherlands, India, Singapore and Switzerland.

A Bioinformatician from Switzerland confirmed that `ipdfc` provides a significant benefit in its simplicity in implementation and accuracy on extrapolating individual-level data. This statistical package can potentially help clinicians with decision making for health technology assessments by the re-analysis of trials, evaluating the benefit of treatments, and enhancing communication with patients. A Data Scientist from Flatz Lab, said "I would like to emphasize that `ipdfc` is exceptionally accurate in our experience... . In summary, using your statistical package has provided me with significant benefits. It has enabled [us] to reduce the time of data collection from months and years to a few minutes and at zero financial cost".

A Spanish hospital confirmed that the `ipdfc` package has enabled their team to reconstruct a database for five studies of 4,686 patients, producing evidence of the long-term survival benefit of a mechanical valve compared to biological aortic valve replacement in patients with heart diseases. Their previous methodology frequently produced incorrect conclusions and Wei's package enabled them to strengthen their methodology. "The package has contributed to producing reliable information which we are confident in, and I found it is more accurate than the other packages for regenerating the original survival data. Your work has been very beneficial... allowing us to increase our understanding of the research questions. The package has informed and benefitted our research practice enabling us to perform alternative statistical analyses", said a clinician from Central University Hospital of Asturias, Spain.

4 Restricted Mean Survival Time

4.1 Methodology

The restricted mean survival time (RMST) [5], μ^* at time t^* of a random variable T, which denotes the time to an event, is defined as

$$\mu^* = \int_0^{t^*} S(t)dt, \tag{2}$$

where $S(t)$ is the survival function for T. The measure μ^* increases monotonically with t^*, because equation (2) gives a non-negative, increasing function of t^*. The RMST can be obtained by using one of the following methods:

1. Pseudo-values with $S(t)$ estimated by a Kaplan-Meier estimate.
2. Flexible parametric survival model where $S(t)$ is estimated by $\exp(-H(t))$ with the baseline log cumulative hazard function modelled by restricted cubic spline.
3. Approximate the integral in equation (2) by the following summation,

$$\sum_{j=1,0<t\leq t^*}^{k} \hat{S}(t_j)(t_{j+1} - t_j) + (t_1 - 0),$$

where $\hat{S}(t_j)$ is the Kaplan-Meier estimate at time $t_j (0 < t \leq t^*)$, t_j is the time where an event occurs and k is the total number of events.

4.2 Application to Survival Outcomes Following Kidney Transplant

Since 2015, Wei conducted statistical analysis for the annual reports of the South West Transplant Centre (SWTC). Wei estimated the RMST [5] between living and deceased donor recipient groups for kidney graft survival and patient survival, respectively, and estimated the difference in RMST between these two groups to give an interpretation of the gain/loss in life expectancy in time scale.

4.3 Impact on Enhancing Communication of Survival Outcomes in Kidney Transplant

Wei's research in kidney transplant survival analysis provided enhanced information, supporting treatment planning and enabling doctors' communication with patients in a more intuitive way. "Thank you for your contributions to the statistical analysis for the kidney transplant data... 2015, 2016, 2017, 2018. This has helped us for the audit of survival outcomes of patients following kidney transplant... [this] is of interest to clinicians and will provide a new piece of medical information which will be helpful in communications with patients", Director, the SWTC.

In 2019, Wei and her team presented their survival analysis at the SWTC annual transplant day. The event was attended by 57 participants, including clinicians, nurses, statisticians, biomedical scientists, pharmacists and patient representatives. A University of Plymouth commissioned independent evaluation of the event confirmed that doctors find Wei's application valuable. One doctor said, "As a clinician, I feel that RMST is very intuitive and easier to explain/more relevant to patients/practice". Seventy-one percent of doctors said that the use of RMST had enabled them to explain years/m"onths gain/loss in life expectancy since transplant, and 67% found RMST easier to understand than the hazard ratio. Visualisation of the difference in RMST helped doctors to understand the change of survival benefit of transplants over time (71%), and 42% of doctors reported the RMST assisted with decision-making about treatment options in their clinical practice. A clinician, University Hospitals Plymouth NHS Trust, said "Your presentation... received extremely positive feedback. Of the participants that responded, 92% said that they found it useful".

5 Conclusion

Through three examples, we have demonstrated the respective impact of our novel medical statistics research with a national and international reach in informing healthy eating guidelines, regenerating individual participant data for health research and technology assessments, and improved clinical practice through enhancing interpretation of survival outcomes.

Acknowledgements YW received funding from EPSRC Impact Acceleration Account (EP/X525789/1).

References

1. Wei, Y., Higgins, J.P.T. (2013). Bayesian multivariate meta-analysis with multiple outcomes. *Statistics in Medicine*, 32(17): 2911–2934.
2. Hollands, G.J., Shemilt, I., Marteau, T.M., Jebb, S.A., Lewis, H.B., Wei, Y., Higgins, J.P.T., and Ogilvie, D. (2015). Portion, package or tableware size for changing selection and consumption of food, alcohol and tobacco. *Cochrane Database of Systematic Reviews*, **9**.
3. Wei, Y., Royston, P. (2017). Reconstructing individual-level time-to-event data using published Kaplan-Meier curves. *The Stata Journal*, 17(4): 786–802.
4. Fox, K.M., EURopean trial On reduction of cardiac events with Perindopril in stable coronary Artery diseases investigators. (2003). Efficacy of perindopril in reduction of cardiovascular events among patients with stable coronary artery disease: randomised, double-blind, placebo-controlled, multicentre trial (the EUROPA study). *Lancet*, 362, 782–788.
5. Wei, Y., Royston, P., Tierney, J.F. and Parmar, M.K. (2015). Meta-analysis of time-to-event outcomes from randomized clinical trials using restricted mean survival time: application to individual participant data. *Statistics in Medicine*, 34(21): 2881–2898.

Improving the Safety and Quality of Healthcare Using Statistical Analysis of Routinely Collected Data

Alex Bottle and Paul Aylin

Abstract The quality of healthcare can be measured in many ways, from staffing to survival rates. England's National Health Service is highly regarded worldwide, but scandals led to increased scrutiny and calls for greater monitoring. Our involvement in two public inquiries led to us develop a national hospital mortality monitoring system using routinely collected inpatient data. Our commercial partners designed the system's web-based front end to alert hospitals when their death rates were significantly above the national average. We devised the technical details of this system, which included how to define clinically important patient groups, how to adjust for patient risk, and how to limit the false alarm rate when monitoring hundreds of patient groups and hospitals over time. We derived an equation to tailor the alerting threshold to the desired false alarm rate, the size of the hospital, and the death rate for each patient group. Our system immediately flagged the high death rate at Mid Staffordshire NHS Trust. It transformed Dr Foster's business model, their expansion leading to substantial investment from the Department of Health, and our mixed-methods evaluation showed its positive impact on the NHS and patients. It continues to help many hospitals monitor and improve their performance.

1 Introduction

Quality of care can be measured in different ways. A common framework distinguishes between measures of structure (the resources one has), process (what members of the healthcare team do) and outcome (the end result of structures and processes, such as survival). Ultimately what matters is the outcome. The most basic outcome is survival, usually measured in terms of death rates—the proportion of

A. Bottle (✉) · P. Aylin
School of Public Health, Imperial College London, White City Campus, 90 Wood Lane, London W12 0BZ, UK
e-mail: robert.bottle@imperial.ac.uk

P. Aylin
e-mail: p.aylin@imperial.ac.uk

P. J. Aston (ed.), *More UK Success Stories in Industrial Mathematics*,
Mathematics in Industry 42, https://doi.org/10.1007/978-3-031-48683-8_20

patients who died within a given timeframe. There are mathematical challenges to measuring and monitoring death rates that we will describe later. The UK is perhaps the world leader in transparency of performance of its National Health Service (NHS). However, one could argue that developments in performance monitoring here have been driven by scandal [1].

Several of these scandals have resulted in public inquiries, led by a judge. We were involved in three of them. Our analysis of routinely collected data on all NHS hospital admissions in England (Hospital Episodes Statistics, HES) for the first of them showed that the death rate for procedures to correct congenital heart abnormalities in babies at Bristol Royal Infirmary was much higher than other specialist centres' rates. Following the inquiry, two surgeons at Bristol performing these operations with poor results despite the concerns of colleagues were struck off the medical register or suspended. Our later analysis showed that the following year, Bristol's death rate for these procedures returned to normal. The next public inquiry followed the eventual arrest of GP Harold Shipman, who was convicted of the murder of 15 of his patients but was suspected of the murder of over 200. Often, he would visit the patient at their home and give them a "tonic", which was in fact diamorphine, better known as heroin. We were asked by the inquiry in 2001 to evaluate available statistical methods for spotting a GP practice with an unusually high death rate. We were given the numbers of patients on GP lists and the numbers of deaths each year, each broken down by broad age group and sex. Using simulated and these real data, we applied a recently published method called the log-likelihood cumulative sum (cusum). Cusums are an example of statistical process control, long used in industry for quality control of manufacturing processes but that had recently been tried in healthcare. We demonstrated that, had this method and available data been used, Shipman would have been flagged at least two years before his arrest. However, this analysis also flagged several other GP practices as high mortality. Subsequent investigation revealed that these practices typically cared for a lot of care home residents, who were older and frailer. It was therefore clear to us that improvements in data were required before a national system would be feasible.

In 2002, the two of us helped found the academic unit at Imperial College London called the Dr Foster Unit, then led by Prof Brian Jarman and largely funded by our commercial partners Dr Foster. At that time the company's main customers were the media, particularly national newspapers. Their main publication was The Good Hospital Guide. Our Unit had and still has access to England's HES data, and we produced analyses on waiting times by specialty and, very controversially and for the first time, hospital death rates. The latter were in the form of Hospital Standardised Mortality Ratios (HSMRs): ratios of the observed deaths to the expected deaths in each hospital. The expected numbers take account of factors such as age and comorbidity (see later section on risk adjustment) so that comparisons are fairer. These were published in the Sunday Times in the form of league tables in the early years. However, there are some key drawbacks to this approach to be overcome—see the section on defining outliers.

During 2004–5, using knowledge gained during the Bristol and Shipman Inquiries, we worked to develop a mortality monitoring system for hospitals using HES data.

For every admission to an NHS hospital (or for an NHS patient in a private hospital) in England, information is collected among other things on age, sex, diagnoses, procedures, and outcome such as death in hospital. There is a de-identified patient ID to allow linkage between records belonging to the same person, which allows readmissions to be defined. Our initial Excel spreadsheet output was turned into an interactive online system by Steve Middleton, chief software engineer at Dr Foster. This system allowed users - staff at hospitals that bought access - to run analyses in real time to show how their rates of death, long length of stay and emergency readmission compared with the national average and with their peer hospitals. Data are presented in tables and graphs on HSMRs, also broken down by diagnosis and procedure, and cusums. This rapidly proved popular within the NHS. At its peak, around 70% of England's hospitals were using the system, then called Real Time Monitoring (RTM); a version of it still runs today under a different name. RTM had a notable early success. In its first year, we sent a number of letters (all unanswered) showing cusum alerts for different patient groups to the same hospital, Mid Staffordshire NHS Foundation Trust; we also sent copies of the letters to the national healthcare regulator, now called the Care Quality Commission (CQC). They also had very high overall mortality as measured by the HSMR published by Dr Foster and the national newspapers. The Trust had laid off nursing and other staff apparently to save money, with terrible consequences [2]. A combination of the public campaign by the relatives of many patients who died at the hospital, concerns by the regulator and our mortality alerts led to a public inquiry in late 2010. This followed a narrower independent inquiry and uncovered many shocking failings in care standards, leading to 290 recommendations [2]. One of these was that "All healthcare provider organisations should develop and publish real time information on the performance of their consultants and specialist teams in relation to mortality. . . .". Our system has since helped many hospitals monitor and improve their patient outcomes (see the section on impact). We now describe the technical details.

2 Risk Adjustment

To allow for fairer comparisons between hospitals, risk adjustment is needed. This is the statistical adjustment for differences in patient characteristics such as age and medical conditions, which vary between hospitals. Once allowance has been made for random variation and patient factors, any remaining variation between hospitals in their outcomes will be due in part to the quality of care, though in practice, data quality will also contribute. With multiple factors to adjust for, some type of model is needed, such as logistic regression [3]. With hundreds of patient groups to risk-adjust, some automation is desirable, which we showed was as good as manual model-building.

Machine learning methods are of much interest in the related field of risk prediction, though with HES data, we did not find any advantage in using them over logistic regression for risk adjustment. HES has very little on test results and physiological parameters such as blood pressure, and there is notable under-recording of several secondary diagnoses. Risk adjustment will therefore be incomplete, leading to false positives: some hospitals that are flagged with having high mortality will have some patients that are sicker than average whose high risk cannot be fully appreciated by the data and hence by the models. With more detailed data, more complex models are possible, but clinical input is essential.

3 Defining Outliers and Controlling the False Alarm Rate

When is a death rate unacceptably high? In general, quality of care can be assessed against a benchmark such as a clinically meaningful figure, a nationally or locally agreed target, or, in the absence of both, some peer group or national average. There is often little agreement on what the rates of a poor outcome should be with best or standard practice, so the national average is frequently used.

Secondly, it is crucial to account for random variation in the rates. The timing and location of anyone's death is dependent on myriad factors, many of which are not known. This variation means that death rates are often presented with 95% confidence intervals to detect whether the rates are statistically significantly different from the benchmark or consistent with random fluctuation. One of the limitations of league tables as used in sport, in which one team comes first and another last, is that they ignore such fluctuations. Labelling a hospital "the worst" in the country is not justifiable statistically and can have unwarranted financial and reputational implications. One way to avoid this is to use Bayesian rankings, which we managed to get into print in the Sunday Times one year. With this method, we ordered hospitals by the probability that they lie in the top or bottom 10% of hospitals in the country. However, we found that the journalists just made a league table out of them in the accompanying commentary, over which we had no editorial control.

Thirdly, any statistical test has a certain level of false positive results or false alarms. The number of false positives increases with the number of tests performed. While one hospital can be compared with the national average, yielding a 95% confidence interval and a p value to indicate statistical significance, with many hospitals and many patient groups within each of those hospitals being similarly compared, the chance of a false positive is very high. This needs to be controlled or faith in the system will be lost, like the boy who cried "wolf". Within RTM, this can be done by changing the threshold at which an alarm is generated. In a cusum plot, one admission is plotted after another during a given period of time, and the chart is updated each month with new data (Fig. 1).

A function of the difference between the observed outcome (0 or 1 for death) and the predicted risk of death for that admission is plotted. The chart will trend upwards if there are more deaths than expected and downwards if there are fewer deaths than expected. If the chart crosses the pre-set threshold, an alarm or "signal" is generated, suggesting that the hospital's death rate is higher than the average. In Fig. 1, this occurs sometime in January. This should trigger an investigation at the hospital to

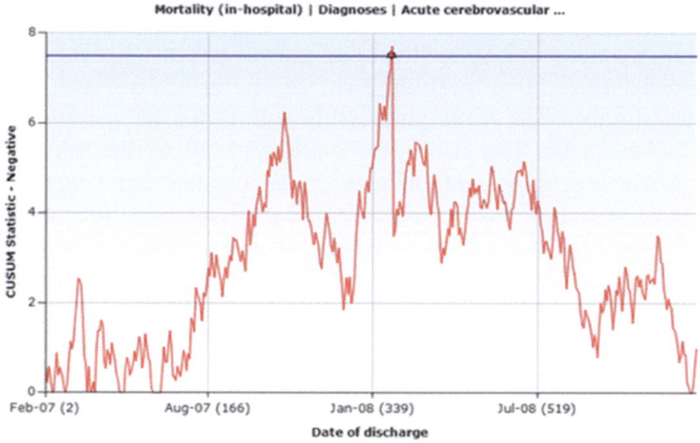

Fig. 1 Example of a cusum plot

determine the cause(s). The threshold can be raised to limit the false alarms, but if the threshold is set too high, the chance of detecting genuine problems with mortality will be too low. There is thereby a trade-off here. The question is where precisely to set the chart threshold to give a desirable false alarm rate. Using Monte Carlo simulation, we derived an equation so that the threshold could be appropriately tailored to the hospital's size and the patient group as well as the desired false positive rate [4]. The false positive rate is set by the user after they log in:

$$\text{False alarm rate} = atp + btp^2$$

where t is the number of admissions, p is the mortality rate, $a = ce^{-dh}$ and $b = -fg^h$, with c, d, f and g being constants, and h being the cusum chart threshold.

4 Impact

Before RTM, the main customer for Dr Foster was the print media, often with TV and radio coverage. Success of early sales of RTM in particular and some other health analytics products transformed the company's business model. The company's growth soon attracted £12 million investment from what was then the Health and Social Care Information Centre through the "joint venture" in 2006. The Health Minister Lord Warner announced at its launch that the venture would create an information market in which the key beneficiaries would be the public. Competition in this nascent market grew over time, and Telstra Health acquired Dr Foster in 2015, reportedly for 25–30 million Australian dollars. NHS budget restraints in recent years have made healthcare organisations more reluctant to spend money on performance information. However, the whole healthcare analytics market is now worth some $3.4bn in Europe alone in 2022, with strong growth forecast, according to a report by Market Data.

The impact on the NHS of our partnership with Dr Foster, particularly through RTM and its successors, has also been considerable. Between August 2013 and December 2020, we sent out 1,312 alert letters to hospitals, covering nearly 41,000 deaths. Various hospitals showed improvements to outcomes after using the tool and/or other mortality figures from Dr Foster. Some examples can be found in hospitals' Quality Accounts documents online, which use data from RTM and other sources to set performance targets for the following years. An example is sepsis at Oxford University Hospital in 2017/8, with reports of improvements in both care processes and mortality for sepsis by the following year.

In 2018, we conducted a mixed methods evaluation of RTM [5]. One part consisted of an analysis of 532 cusum alerts sent to 139 hospital Trusts. Each alert is accompanied by a letter to the CQC, who often follow up with the hospital, requiring them to investigate the reasons for the alert. We analysed the CQC's documents of these investigations and found that quality of care was cited as a factor in some 70% of them. We also conducted a national survey and site visits to 11 hospitals. Full action plans for improvement were created in 77% of trusts investigated, and several hospitals reported positive changes in culture and processes, with subsequent benefits to patients.

RTM is still in use today under a different name and with further functionality. Our work has helped drive greater transparency and focus on outcomes. The market for healthcare analytics continues to grow globally. In the future, one can expect to see advances in risk-adjustment methods following improved but safe access to more detailed and timely patient data in the form of electronic health records, further reducing false positives in tools such as RTM. The blend of statistical knowledge, experience and rigour from academia with innovation, design and market reach from industry has made a positive impact on patient care and business for over 15 years—and counting.

Acknowledgements The authors are grateful for the financial support in the form of research grants to the Unit between 2002 and 2021 and to NIHR for funding our mixed methods evaluation of RTM.

References

1. Bottle, A. and Aylin, P.: Statistical Methods for Healthcare Performance Measurement. Taylor and Francis, London (2016)
2. Mid Staffordshire NHS Foundation Trust Public Inquiry. Report of the Mid Staffordshire NHS Foundation Trust Public Inquiry. London: The Stationary Office (2013).
3. Bottle, A. and Aylin, P.: Intelligent information: a national system for monitoring clinical performance. Health Serv. Res. **43**, 10–31 (2008).

4. Bottle, A. and Aylin, P.: Predicting the false alarm rate in multi-institution mortality monitoring. J. Operational Res. Soc. **62**, 1711–1718 (2011).
5. Aylin, P., Bottle, A., Burnet,t S., Cecil, E., Charles, K.L., Dawson, P., D'Lima, D., Esmail, A., Vincent, C., Wilkinson, S., Benn, J.: Evaluation of a national surveillance system for mortality alerts: a mixed-methods study. Health Serv. Delivery Res. **6**, 7 (2018).

Pharmaceuticals

Minimising Bias and Increasing Relevance in Clinical Trials: Basis for the International Standard

Paula R. Williamson and J. L. Hutton

Abstract Worldwide clinical trials are the mechanism by which the safety and effectiveness of medicines and interventions are assessed. The effectiveness of interventions are assessed by outcomes such as decreased blood pressure or frequency of seizures, or changed weight. Harms of treatments are also relevant. Statisticians demonstrated mathematically that there was a systemic problem with outcome selection and reporting bias across the entire clinical trials sector. The impacts of taking multiple measurements and only reporting a selected subset of measurements, or selected subgroups of patients were modelled. This provided clear evidence of the need for change. This work led to the COMET Initiative (Core Outcome Measures in Effectiveness Trials), with important policy and societal consequences. COMET is now the international standard for harmonising outcomes for specific health conditions. Decisions about treatments are made more robust by ensuring that effectiveness data can be compared and combined. Trials have become more relevant to patients. COMET has transformed patient participation in outcome-setting from 16% up to 2013 to 93% in 2020. COMET is endorsed by mutliple healthcare decision-making organisations including guideline developers such as NICE, and trial funders including the NIHR and the pharmaceutical industry.

1 The Problem of Biased Reporting in Medical Journals

Worldwide, clinical trials are the mechanism by which the safety and effectiveness of medicines and interventions are assessed, by comparing relative effects on health outcomes. Results from clinical trials guide treatment decisions for patients and health professionals. Clinical trials are the gold standard for evaluating treatments such as drugs, procedures or psychological interventions. Further, systematic reviews

P. R. Williamson
Department of Health Data Science, University of Liverpool, Liverpool L69 3GF, UK
e-mail: p.r.williamson@liverpool.ac.uk

J. L. Hutton (✉)
Department of Statistics, University of Warwick, Coventry CV4 7AL, UK
e-mail: J.L.Hutton@warwick.ac.uk

P. J. Aston (ed.), *More UK Success Stories in Industrial Mathematics*,
Mathematics in Industry 42, https://doi.org/10.1007/978-3-031-48683-8_21

and meta-analysis of clinical trials enable the results from different trials to be compared, combined and contrasted, for better informed decisions about treatments, and to robustly establish which treatments work.

The problem of publication bias, whereby a trial may be published or not depending on the results, has previously been identified and is a potential threat to the appropriate interpretation of evidence. Initiatives were implemented, requiring a trial to be registered and results included in a trials registry, thus increasing the ability to find trials undertaken.

A further form of selective reporting bias within trials was more recently uncovered. Thousands of different outcomes may be identified across many different diseases and conditions, with a lack of consistency in outcomes selected for measurement across trials for particular conditions. A form of within-study reporting bias, in which outcomes within in a particular trial are reported or not in the publication depending on the results, was subsequently exposed. Empirical evidence for the widespread existence of this problem, known as outcome reporting bias (ORB), exists. The impact of ORB has been shown to be significant, overestimating the treatment effect by 20% in a third of systematic reviews. Combining this lack of consistency in outcomes selected for measurement across trials for particular conditions, and subsequent selective reporting of outcome results in a biased way, make comparing and combining trial data difficult [5].

Professors Williamson and Hutton collaborated to propose statistical methods to assess the potential consequences of selective outcome reporting bias in clinical trials, and to establish the robustness of a trial, and subsequent meta-analyses, to this problem.

2 Selective Reporting

Outcomes from trials can be selected for inclusion in published research articles in two main ways. Several outcomes might be measured, and authors report only those which they consider interesting [4].

2.1 Selective Reporting of Multiple Outcomes

Suppose that researchers calculate test statistics for all the outcomes which have been recorded. They then report the outcome with the smallest significance level. It is common in meta-analysis to invoke the central limit theorem, and thus to base estimation for binary, ordinal and survival data, as well as continuous data, on the Gaussian distribution. We therefore use the multivariate Gaussian (normal) distribution to model selection. The choice of the smallest significance level can be modelled as taking the largest order statistic on the effect size scale, which is defined as the ratio of the mean to the standard deviation. Consider a trial with p outcomes, $X_i, i \in 1, \ldots, p$,

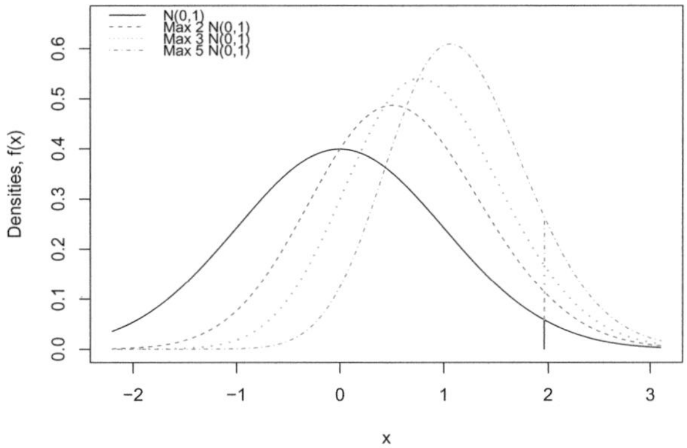

Fig. 1 Density functions for maximum of 1, 2, 3 and 5 standard normal variates

with all outcomes independently and identically distributed (IID) $N(\mu, \sigma^2)$, which reports the maximum of p $N(\mu/\sigma, 1)$, statistics. The probability density function of $Y_{(p)}$, the maximum of p IID $N(\mu, \sigma^2)$ variates, is

$$f_{Y_{(p)}}(y) = p\sigma^{-1}\Phi\left(\frac{y-\mu}{\sigma}\right)^{p-1}\phi\left(\frac{y-\mu}{\sigma}\right), \tag{1}$$

where $\Phi(.)$ and $\phi(.)$ denote the standard normal distribution and density functions.

Figure 1 illustrates the area under the curve to the right of the 97.5% quantile of the standard normal distribution, 1.96, the basis of a two-sided significance test, and the effect of selecting from several outcomes. The rapid increase in the area under the curve is clear. Table 1 show the increase significance levels for a single trial with different numbers of outcomes.

Table 1 Selection of the maximum of two, three or five correlated outcomes
Upper 97.5% point for within-trial selection; $F(.)$ and effect size bias with multiple trials

No. of trials	Outcomes	Statistic	Correlation ρ between N(0,1) variates				
			$\rho = 0.9$	$\rho = 0.7$	$\rho = 0.5$	$\rho = 0.1$	$\rho = 0.0$
One	Two	$F^{-1}(0.975)$	2.108	2.180	2.212	2.237	2.239
One	Three	$F^{-1}(0.975)$	2.185	2.298	2.349	2.388	2.391
One	Five	$F^{-1}(0.975)$	2.274	2.436	2.512	2.568	2.573
Two	Two	$P(y > 1.96)$	0.06	0.07	0.07	0.08	0.08
Two	Two	Bias	0.52	0.57	0.63	0.76	0.80
Three	Two	$P(y > 1.96)$	0.08	0.09	0.09	0.11	0.12
Three	Two	Bias	0.65	0.71	0.77	0.94	0.98

2.2 Systematic Reviews and Meta-Analyses

A major advance in health care at the end of the 20th century was the use of systematic reviews to inform treatment decisions. The impact of selective reporting on the validity of systematic reviews and meta-analyses is relevant to decision-making [4, 5]. Even with selection from only two independent outcomes, from two or three trials, the bias in the estimated effect size is substantial, 0.8 or 0.98 Table 1. Of course, it is not likely that the outcomes will all be independent. The actual significant levels or p-values for a result above 1.96 given in Table 1 for a range of correlations between variables show that even a strong correlation of 0.9 more than doubles the p-value.

2.3 Selective Non-reporting of Multiple Outcomes

These reviews can be affected both by selective reporting of outcomes and and if an outcome is measured but not reported. Evaluation of a large cohort of systematic reviews reported that the majority, 55%, did not include results for the primary review outcome [5]. A tool was developed to score ORB [5]. Treatments almost always have side effects, which are even more open to selective non-reporting. Likelihood-based methods model the effect of ORB, based on the probability that the outcomes are measured, and if measured, the probability of the result being reported [1, 2]. As the assessment of the risk of non-reporting is not perfect, this uncertainty is also modelled by two parameters for reliability. The bias depends on the number of patients in trials with high risk of bias relative to the total number of patients in trials which report the outcome.

3 International Guidelines: Prophylactic Solution

The rigorous study of, and statistical adjustment for the detrimental effects of poor outcome selection and reporting bias within individual clinical trials and meta-analysis motivated Professor Williamson to consider ways to reduce the problem at the trial design stage, through the implementation of core outcome sets (COS). A Core Outcome Set (COS) for specific health condition is an agreed, minimum set of outcomes that should be measured and reported in all clinical trials undertaken for that condition. A COS ensures that outcomes relevant to decision-makers, including patients, are recorded and reported.

Prof Williamson and colleagues founded the COMET (**C**ore **O**utcome **M**easures in **E**ffectiveness **T**rials) Initiative, www.comet-initiative.org, in 2010 to improve the consistency of outcome selection and reporting across the whole clinical trials sector. In 2014 they populated the COMET database with all available COS. This systematic review of COS has since been updated annually - the database currently contains over 500 published and more than 400 registered ongoing studies - to ensure COMET resources remain current.

This work allowed COMET to show that up to August 2013 only 16% (31/198) of all published COS involved members of the public. Through its People and Patient Participation (PoPPIE) Group, COMET has established a patient voice in COS development that did not previously exist, and made this standard practice across the clinical trials sector. Over 90% of all COS currently in development now include patients and their representatives as participants, ensuring their perspectives are central to decisions about what outcomes should be core for their health condition.

COMET has been adopted by key healthcare research and decision-making organisations, including the National Institute for Health and Care Excellence (NICE) and other regulators, funders and pharmaceutical companies in the UK, Europe and the US, https://www.comet-initiative.org/COSEndorsement:

1. Increasing numbers of public research funding organizations, including the National Institute for Health and Care Research (NIHR) and PCORI in the US, are including formal guidance to applicants to consider a COS in their study and include it unless there is a good reason to do otherwise.
2. The UK Health Research Authority best practice guidance for clinical trial design sets out '*an expectation that the core outcomes will be collected and reported*' using COMET.
3. The UK's main trial registry ISRCTN, recognised by the World Health Organisation (WHO), states users should consult COMET before registering their study.
4. COMET is enabling the pharmaceutical industry to select and report consistent outcomes for commercially funded clinical trials. The *Big Data for Better Outcomes* toolkit was developed by a consortium of the world's largest pharmaceutical companies, citing the COMET database throughout, as best practice in commercial trials.
5. COMET is improving recommendations for clinical practice. NICE has incorporated COMET's resources into its manuals for clinical guideline development and health technology evaluations. NICE also uses COMET to identify outcomes for use in NICE Quality Standards, helping to ensure measures that '*matter most for patients and service users*'.

COMET is now the international standard for harmonising outcomes - making decisions about treatments more robust by ensuring that effectiveness data can be compared and combined, and making trials relevant to patients.

4 International Guidelines: Wider Relevance

Whether demographic or disease specific factors affect outcomes is often an important consideration. Issues arising from reporting of selected results for only statistically significant differences between sub-groups such as sex or disease severity [3] could be addressed in future guidelines.

The issues which are addressed here are not specific to health research. Outcome reporting bias is recognised as a problem in research articles in education and pyschology. Social scientists typically regard effect sizes of 0.5 as medium, which is similar to the bias resulting from selecting the more significant of two moderately correlated variables to report. Such biases are likely to be an important aspect of the reproducibility crises.

Acknowledgements Prof Williamson received funding from the MRC (MR/J004847/1 and MR/K025635/1) and the European Commission FP7 (305081) for the set up and establishment of COMET.

References

1. J Copas, KM Dwan, JJ Kirkham, and PR Williamson. A model-based correction for outcome reporting bias in meta-analysis. *Biostatistics*, 15(2):370–383, 2013.
2. J Copas, AG Marson, PR Williamson, and JJ Kirkham. Model-based sensitivity analysis for outcome reporting bias in the meta analysis of benefit and harm outcomes. *Statistical Methods in Medical Research*, 28(3):889–903, 2019.
3. S Hahn, P R Williamson, J L Hutton, P Garner, and V Flynn. Assessing the potental for bias in meta-analysis due to selective reporting of subgroup analyses within studies. 19:3325–3336, 2000.
4. J L Hutton and P R Williamson. Bias in meta-analysis with variable selection within studies. 49:359–370, 2000.
5. JJ Kirkham, KM Dwan, DA Altman, C Gamble, S Dodd, R Smyth, and PR Williamson. The impact of outcome reporting bias in randomised controlled trials on a cohort of systematic reviews. *BMJ*, 340, 2010.

Using Mathematical Models of Heart Cells to Assess the Safety of New Pharmaceutical Drugs

Gary R. Mirams

Abstract Many drugs have been withdrawn from the market worldwide, at a cost of billions of dollars, because of patient fatalities due to drugs unexpectedly disturbing heart rhythm. Even drugs for ailments as mild as hay fever have been withdrawn due to an unacceptable increase in risk of these heart rhythm disturbances. Consequently, the whole pharmaceutical industry expends a huge effort in checking all new drugs for any unwanted side effects on the heart. The predominant root cause has been identified as drug molecules blocking ionic current flows in the heart. Block of individual types of ionic currents can now be measured experimentally at an early stage of drug development, and this is the standard screening approach for a number of ion currents in many large pharmaceutical companies. However, clinical risk is a complex function of the degree of block of many different types of cardiac ion currents, and this is difficult to understand by looking at results of these screens independently. By using ordinary differential equation models for the electrical activity of heart cells (electrophysiology models) we can integrate information from different types of currents, to predict the effect on whole heart cells and subsequent risk of side effects. The resulting simulations can provide a more accurate summary of the risk of a drug earlier in development and hence more cheaply than the pre-existing approaches.

1 Introduction

The contraction of your heart is both triggered and co-ordinated by a wave of electrical activity passing through each of its muscle cells. This wave propagates because ions of various types—particularly sodium, potassium and calcium—move into and out of each cell on every beat. The ions move through thousands of protein complexes that are in each cell's membrane. Some of these protein complexes passively allow

G. R. Mirams (✉)
School of Mathematical Sciences, Centre for Mathematical Medicine & Biology, University Park, Nottingham NG7 2RD, UK
e-mail: gary.mirams@nottingham.ac.uk

P. J. Aston (ed.), *More UK Success Stories in Industrial Mathematics*,
Mathematics in Industry 42, https://doi.org/10.1007/978-3-031-48683-8_22

specific types of ions to flow along their electrochemical gradient (this would be simply from high to low concentration but there is also a contribution from the electric field), these are called *ion channels*. Other complexes are *ion pumps*, which use energy to actively move an ion against its gradient; and another type are *ion exchangers*, which move one type of ion against its gradient using the energy from the flow of another type down its gradient. These channels, pumps and exchangers are encoded by particular genes, and typically we consider 10–20 genes and their associated types of current carriers to be the most important in explaining the heart's electrical activity. Many of the currents are selective for just particular ions, and they respond to the trans-membrane voltage of the cell in different ways (opening or closing, rapidly or slowly, at different voltages).

Some of the heart's ion channels are particularly prone to being blocked by many pharmaceutical drug compounds. Sometimes this is the desired action of a drug, to treat disturbances to healthy heart rhythm which are known as *arrhythmias*. But unfortunately many drugs (even for diseases unrelated to the heart) also block cardiac ion channels as a side effect; this can have serious consequences in terms of increasing the risk of patients suffering fatal arrhythmias. As a result large amounts of time and money are now spent during the drug development process to test whether potential drug compounds are likely to have these pro-arrhythmic side effects on the heart.

In the first instance, pharmaceutical companies can check whether new compounds interact with cardiac ion channels in automated experiments for one type of ion channel at a time. This 'screening' effort has been relatively successful; very few compounds reach the market and then have unexpected effects on the heart. Regulators of the pharmaceutical industry, such as the USA's Food and Drug Administration, have become more concerned that we may be discarding potentially good drug compounds that have small effects on the heart which would not actually be pro-arrhythmic. So in recent years we have been working with pharmaceutical companies and international regulators to run electrophysiology simulations using mathematical models to improve and refine the assessment of pro-arrhythmic risk of novel drug compounds.

2 Modelling the Cardiac Action Potential

The fundamental model for electrical activity of cells is relatively simple: the voltage across the cell membrane (V) changes as a result of ionic currents flowing across the membrane. In ordinary differential equation (ODE) form, this reads

$$\frac{\mathrm{d}V}{\mathrm{d}t} = -\frac{1}{C_m}\sum_{i}^{N} I_i, \tag{1}$$

where C_m is the cell capacitance, and there are N different types of ionic membrane currents, each denoted by I_i. We show an example model in Fig. 1A.

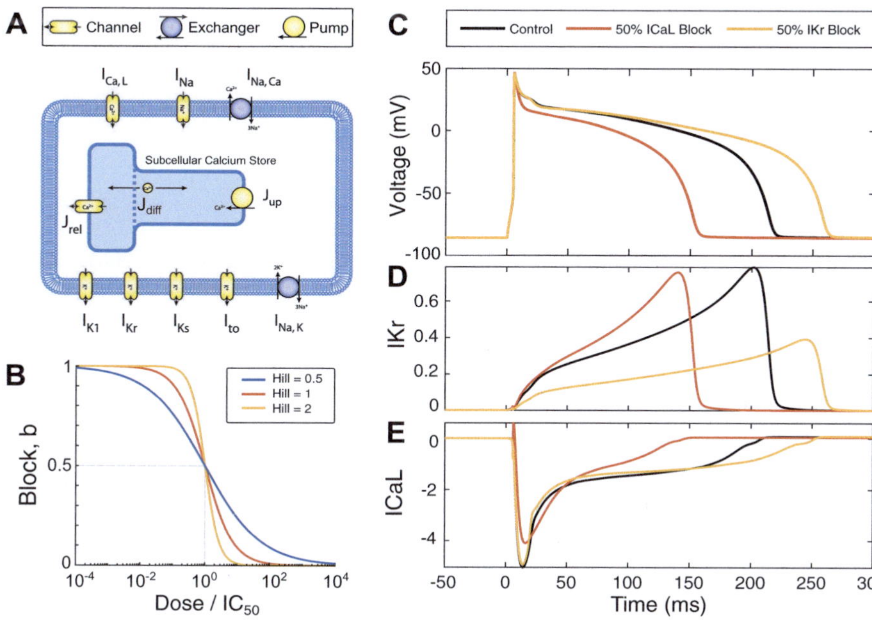

Fig. 1 Models for the electrical activity of cardiac cells and drug block of ion currents. **A** A schematic of a cardiac electrophysiology model showing the predominant ionic currents and the sub-cellular calcium store (sarcoplasmic reticulum), adapted from www.cellml.org. The voltage across the cell membrane will change in response to the sum of all the currents crossing the outer membrane, following Eq. (1). **B** The Hill equation shown with the ratio of dose and IC_{50} on the x-axis, varying Hill coefficients (n in Eq. (7)) shown. A simulation using the Shannon 2004 rabbit ventricle model (https://bit.ly/shannonrabbit) showing the effects of drug block of I_{Kr} or I_{CaL} by 50% on: **C** the membrane voltage/action potential waveform; **D** IKr, an outward repolarising current; and **E** ICaL, an inward current (currents shown in μA/cm^2)

This becomes an interesting and complex feedback model because the currents I_i themselves evolve according to more ODEs that are nonlinear functions of V.

The simplest ion channel current models typically look like

$$I_i = g_i \cdot \mathcal{P}_i(V) \cdot (V - E_i), \tag{2}$$

where $\mathcal{P}_i$ is the open probability of the channels of a given type i, g_i is a parameter representing the maximal current that can flow when these channels are open (considered proportional to the density of channels in the membrane), and E_i is the Nernst/reversal potential for this type of ion. The open probability $\mathcal{P}_i$ is a function of V, and this is usually represented with further sub-systems of ODEs, describing how rapidly this particular current's channels open and close in response to the applied voltage (V).

3 Modelling Drug Action on Ion Channels

We can modify Eq. (2) to include a scaling factor, b, that represents the proportion of current remaining after block with a drug at concentration [D]

$$I = g \cdot b([\mathrm{D}]) \cdot \mathcal{P}(V) \cdot (V - E). \tag{3}$$

To derive an expression for b as a function of drug concentration we consider the following reversible chemical reaction: n drug compounds (D) bind to one ion channel (C) to produce a complex (CD_n) where the ion channel is blocked, at forward rate k_{on} and backward rate k_{off}:

$$\mathrm{C} + \mathrm{nD} \underset{k_{\mathrm{off}}}{\overset{k_{\mathrm{on}}}{\rightleftharpoons}} \mathrm{CD_n}. \tag{4}$$

Assuming mass-action kinetics (reaction rates are directly proportional to the product of their reactant concentrations) allows us to formulate an ODE for C:

$$\frac{\mathrm{d}[\mathrm{C}]}{\mathrm{d}t} = -k_{\mathrm{on}}[\mathrm{C}][\mathrm{D}]^n + k_{\mathrm{off}}[\mathrm{CD}_n], \tag{5}$$

where square brackets denote concentrations. We can divide through by the total number of channels (a constant, $[\mathrm{C}] + [\mathrm{CD}_n]$) to work with b, defined as the proportion of unblocked channels as above. At steady state we get

$$b_\infty([\mathrm{D}]) = \frac{\frac{k_{\mathrm{off}}}{k_{\mathrm{on}}}}{[\mathrm{D}]^n + \frac{k_{\mathrm{off}}}{k_{\mathrm{on}}}}. \tag{6}$$

At 50% block we will have $[\mathrm{D}]^n = \frac{k_{\mathrm{off}}}{k_{\mathrm{on}}}$; the concentration at which this happens is called the '50% Inhibitory Concentration' or $[\mathrm{IC}_{50}]$, so that $[\mathrm{IC}_{50}]^n = \frac{k_{\mathrm{off}}}{k_{\mathrm{on}}}$. We can substitute this in to get

$$b_\infty([\mathrm{D}]) = \frac{[\mathrm{IC}_{50}]^n}{[\mathrm{D}]^n + [\mathrm{IC}_{50}]^n} = \frac{1}{1 + \left(\frac{[\mathrm{D}]}{[\mathrm{IC}_{50}]}\right)^n}. \tag{7}$$

This equation is known as the Hill equation. We can substitute it in place of b within Eq. (3) to simulate the steady-state block of a current as a function of drug concentration, which is then parameterised by the IC_{50} and n which is known as the 'Hill coefficient'. In Fig. 1B we show plots of Eq. (7) for three values of the Hill coefficient n.

Figure 1C shows how the cell's membrane voltage changes in response to $b_\infty = 0.5$, i.e. 50% block, for either I_{Kr} or I_{CaL}, and the coupled changes to currents in Fig. 1D&E. In reality, different compounds will have different affinities for many different channels, and the models as described above allow us to predict the whole-cell effect on electrophysiology.

4 Evaluating Model Predictions

In [1] we used the approach outlined above for multiple ion channels (I_{Kr}, I_{CaL}, I_{Na}), and predicted the resulting action potential duration (length of the wave in Fig. 1C). We showed how classifying drug risk based on this simulation output was more predictive of clinical risk of arrhythmia than simply considering the I_{Kr} IC_{50} or safety margin (concentration of drug reached in clinical dosing/IC_{50}), which had been the predominant strategy up to that point.

Another study tested model predictions for the effects of hundreds of compounds on rabbit heart tissue that had been measured over many years as part of drug development at GlaxoSmithKline Plc. [2]. This study showed that simulations could reach useful levels of predictive power across a whole portfolio of compounds under development within the pharmaceutical industry, and highlighted the accuracy of different pharmaceutical screening technologies.

Because there is uncertainty in the experimental data that the simulations are based upon it is important to consider the consequences of this on model outputs. As shown in Fig. 2 and discussed in its caption, the first part of this task is 'uncertainty

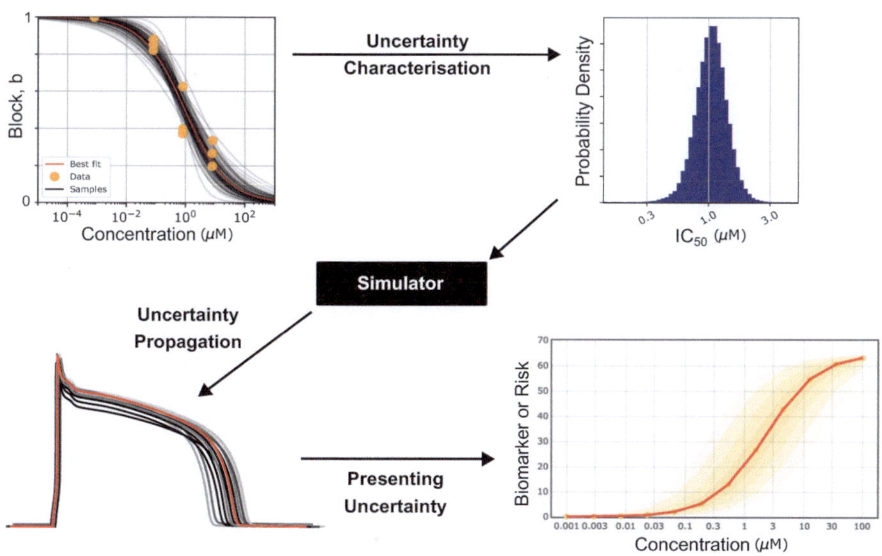

Fig. 2 Uncertainty Quantification for simulations of drug action on electrophysiology. For each ion current that is screened some concentration/block data points are measured. There is resulting uncertainty in the underlying compound properties such as IC_{50}, that we can estimate with an 'uncertainty characterisation' procedure such as Bayesian inference for the parameters of interest (IC_{50} shown here in top right, with corresponding sampled Hill curves in the top left). The simulator then takes as an input a probability distribution, rather than a single number, and produces a corresponding distribution of outputs. The end-user is then presented with a visualisation of the probability that simulation outputs (such as clinical risk) will be in certain ranges, shown here as a fan chart

characterisation'—establishing the uncertainty or variability in the model inputs [3]. The second part of the task is 'uncertainty propagation', mapping the input uncertainty through a simulator to see how this uncertainty affects simulation outputs. This is especially important when predicting clinical risk classes [4] that may decide whether a drug progresses further in development or is approved without restrictions.

These validation studies and a consideration of uncertainty in the simulation results provided enough help with decision making for the pharmaceutical industry and pharmaceutical regulators to pursue the impact discussed below.

5 Impact

A web portal is publicly available (https://cardiac.nottingham.ac.uk) where anyone can register for an account and run cardiac safety simulations by entering data on the degree to which compounds block specific ion channels [5]. As of 2021, this tool has performed over 7000 simulations for over 350 users from 147 companies including Contract Research Organisations, small pharmaceutical companies and 14 of the top 20 largest global pharmaceutical companies; and over 70 educational or research establishments.

The simulator has also been deployed inside company firewalls at GlaxoSmithKline and Roche and is used by Safety Pharmacology teams at both companies for routine safety testing and compound profiling. The simulation results are used to propose concentrations for further tests, check for alignment of later test results with expectations from ion channel screening, and even to replace some animal tissue experiments completely.

The Comprehensive in-vitro Pro-arrhythmia Assay (CiPA, www.cipaproject.org) is a worldwide effort of the pharmaceutical industry, regulators and academia to improve pro-arrhythmic risk assessment for new drugs, and it has included the simulation approaches discussed above. Our work on uncertainty quantification with the US Food and Drug Administration (FDA) led to changes in the proposed clinical risk output from CiPA simulations to make it more robust. The new risk marker was tested in a blinded validation study [4], found to provide excellent predictions, and is now the FDA's primary pre-clinical risk marker to assess the pro-arrhythmic risk of all new pharmaceutical drugs. Pharmaceutical safety testing guidelines are aligned worldwide by the International Council for Harmonisation of Technical Requirements for Pharmaceuticals for Human Use (ICH), and they recently completed a new Question and Answer document to provide guidelines on how to use simulated pro-arrhythmic risk predictions in line with the CiPA approach. Pharmaceutical companies and regulators worldwide can now use the simulation approaches discussed above to assess the safety of new compounds.

Acknowledgements The author gratefully acknowledges support from the Wellcome Trust (grant number 212203/Z/18/Z). He would also like to acknowledge helpful discussions with his research team based at the University of Nottingham, collaborators and members of the Comprehensive in-vitro Proarrhythmia Assay (CiPA) Steering Committee and working groups.

References

1. Gary R Mirams, Yi Cui, Anna Sher, Martin Fink, Jonathan Cooper, Bronagh M Heath, Nick C McMahon, David J Gavaghan, and Denis Noble. Simulation of multiple ion channel block provides improved early prediction of compounds' clinical torsadogenic risk. *Cardiovascular Research*, 91(1):53–61, 2011.
2. Kylie A Beattie, Chris Luscombe, Geoff Williams, Jordi Munoz-Muriedas, David J. Gavaghan, Yi Cui, and Gary Richard Mirams. Evaluation of an in silico cardiac safety assay: Using ion channel screening data to predict QT interval changes in the rabbit ventricular wedge. *Journal of Pharmacological and Toxicological Methods*, 68(1):88–96, 2013.
3. Ross H Johnstone, Rémi Bardenet, David J Gavaghan, and Gary R Mirams. Hierarchical bayesian inference for ion channel screening dose-response data. *Wellcome Open Research*, 1(6), 2016.
4. Zhihua Li, Bradley J. Ridder, Xiaomei Han, Wendy W. Wu, Jiansong Sheng, Phu N. Tran, Min Wu, Aaron Randolph, Ross H. Johnstone, Gary R. Mirams, Yuri Kuryshev, James Kramer, Caiyun Wu, William J. Crumb, and David G. Strauss. Assessment of an In Silico Mechanistic Model for Proarrhythmia Risk Prediction Under the Ci PA Initiative. *Clinical Pharmacology & Therapeutics*, 105(2):466–475, 2019.
5. Geoff Williams and Gary R Mirams. A web portal for in-silico action potential predictions. *Journal of Pharmacological and Toxicological Methods*, 75:10–16, 2015.

Animal Welfare

Statistically Designed Veterinary Tools to Assess Pain and Quality of Life in Dogs and Cats

E. Marian Scott and Jacqueline Reid

Abstract The assessment of pain and quality of life for companion animals is strongly subjective and the need for psychometrically valid and robust statistically designed tools has been long recognised. These tools are essential in facilitating assessment of analgesic strategies and therapeutic agents to improve animal welfare. Statistical models and experimental design principles have been developed and underpin two clinical tools to assess acute pain and two owner-based tools to assess chronic pain and health related quality of life impacts in dogs and cats respectively.

1 Introduction

Pain is a vital sign and veterinarians widely recognise its central role in animal care and welfare. However, pain assessment in companion animals is challenging due to the inherent barriers to communication, and the species-specific response to and presentation of pain. The UK companion animal population is estimated to include 10.9 million cats and 9.9 million dogs, many of which at some point in their lives may encounter pain through injury or surgery.

How do we begin to assess pain and quality of life? Our approach has been based on psychometric principles, including the development of structured questionnaires, and psychometrically valid scoring models (using item response theory, principal components and factor analysis). Together, these underpin the creation of the tools and statistical research has been instrumental in, demonstrating the tools' validity

E. M. Scott (✉)
School of Mathematics and Statistics, University of Glasgow, Glasgow G12 8QW, UK
e-mail: Marian.Scott@glasgow.ac.uk

J. Reid
School of Biodiversity, One Health and Veterinary Medicine, University of Glasgow, Glasgow G61 1QH, UK
e-mail: Jacqueline.Reid@glasgow.ac.uk

P. J. Aston (ed.), *More UK Success Stories in Industrial Mathematics*,
Mathematics in Industry 42, https://doi.org/10.1007/978-3-031-48683-8_35

and reliability, essential properties for any tool used in clinical practice and in joint decision making between owners and veterinarians and in evaluating intervention levels to support decision making.

2 Acute Pain Tools for Dogs and Cats—Statistical Research

The original Glasgow Composite Measure Pain Scale (CMPS) measured acute pain in dogs using psychometric methodology to a level of precision suitable for clinical trials. The basis of the scale was observations of the dog's behaviour in a controlled environment. Initially, 279 words and expressions suggested by 69 veterinary surgeons were reduced into 47 words and expressions which were allocated into seven behaviour categories: demeanour and response to people, posture, mobility, activity, response to touch, attention to painful area and vocalisation. Three statistical methods, hierarchical agglomerative cluster analysis, Cronbach's alpha coefficient, and analysis of variance with multiple comparisons and empirical cumulative distributions, were used to validate these categories, and a questionnaire accompanied by a list of definitions was designed around the expressions. However, for routine clinical use, where the emphasis is on speed, ease of use, and guidance for analgesia provision, a short form (CMPS-SF) was developed [1]. This resulted in a web-based app using statistical measurement theory to analyse six behavioural categories with associated vocalisation, attention to wound, mobility, response to touch, demeanour and posture/activity. Originally also we made use of a scaling model, so that each expression received a weight, but when the simplified form was created we moved to a simpler scoring system.

The acute pain cat tool (CMPS-Feline) [2], was developed initially using the same statistical approach as for the CMPS (for dogs) but in addition incorporated analysis of facial expressions around the eyes, ears and mouth. Statistical research used spatial, shape and landmark analysis to identify anatomical landmarks that discriminated between pain free and pained cats to deliver the simplified facial elements of the tool [3].

For both scales, there was considerable work in validation of the scales, and with a focus on the communication and interpretability of the tool outputs, we also developed and quantified intervention and alert levels using discriminant analysis techniques, to inform clinical decision making. The resulting measurement scales (for both dogs and cats) have been validated in different medical and surgical conditions (cancer, osteoarthritis, obesity), using statistical methods for optimal design and analysis and are now being used to validate the analgesic efficacy of pharmaceutical products in clinical trials.

3 Chronic Pain and Health Related Quality of Life (HRQL) Tools for Dogs and Cats—Statistical Research

Following similar psychometric principles, a HRQL questionnaire was developed based around observation of the animal by their owner [4]. The core of the questionnaire consisted of 109 simple descriptor items, each with an associated 7-point (0 to 6) Likert-type rating scale (this is a scale on which responses are framed on an agree-disagree continuum). A factor analysis was then used to identify the latent variables, and also a natural scaling model. Four HRQL domains for dogs were identified using the factor analysis, and these are labelled as happy/content, calm/relaxed, energetic/enthusiastic and active/comfortable. The original HRQL tool for dogs was a paper-based assessment which was too long for owners to easily and routinely use. With a focus on the communication and interpretability of the tool outputs, our statistical research demonstrated the validity of a refined 22-item shortened assessment, superior to the original instrument in its ability to distinguish between healthy and unwell animals. The original items were reduced using expert judgment and factor analysis with items being removed on the basis of item loadings and communalities on factors identified through factor analysis of responses from owners of healthy and unwell dogs, intrafactor item correlations, and ability of individual items to discriminate between healthy and unwell dogs. Validity was also assessed through factor analysis and a field trial using a "known groups" approach. Test-retest reliability was assessed using intraclass correlation coefficients. The shortened tool was then validated in different medical and surgical conditions (cancer, osteoarthritis, obesity), using statistical methods for optimal design and analysis (including time trend analysis and analysis of variance). This tool can be accessed via a mobile phone app for UK/Europe veterinary surgeons and pet owners and is also being used to assess the efficacy of analgesic compounds in clinical trials. Subsequent statistical work has been in the creation of population norms for the HRQL dog and cat tools, and in providing alerts for vets and pet owners [5]. The feline generic quality of life tool was developed in a similar manner in 2018 (HRQL-Feline/VetMetrica Feline) co-funded by Newmetrica and a SMART grant from Scottish Enterprise. The cat tool only identified three domains (or latent variables) labelled as vitality, comfort and emotional wellbeing. In some studies, we have asked for owners and veterinarians to provide their own evaluation of the level of pain that an animal is in and also the impact of any condition on their quality of life (both scores are returned on a zero to 10 where 10 is could not be worse). Figures 1 and 2 show some simple graphical output from such a study. Figure 1 shows the distribution of domain scores for a subset of cats while Fig. 2 shows the relationship between the questionnaire score (3 domains) and the owner/vet assessed pain and quality of life score.

An example of the type of statistical model used in deriving the alert levels is a multivariable logistic regression which models the relationship between the four HRQL domain scores as covariates and the health status of the dog (healthy/not-healthy) as the response. Variable selection was carried out using 10-fold cross validation to identify the HRQL domains to be used, based on how well they predicted the health

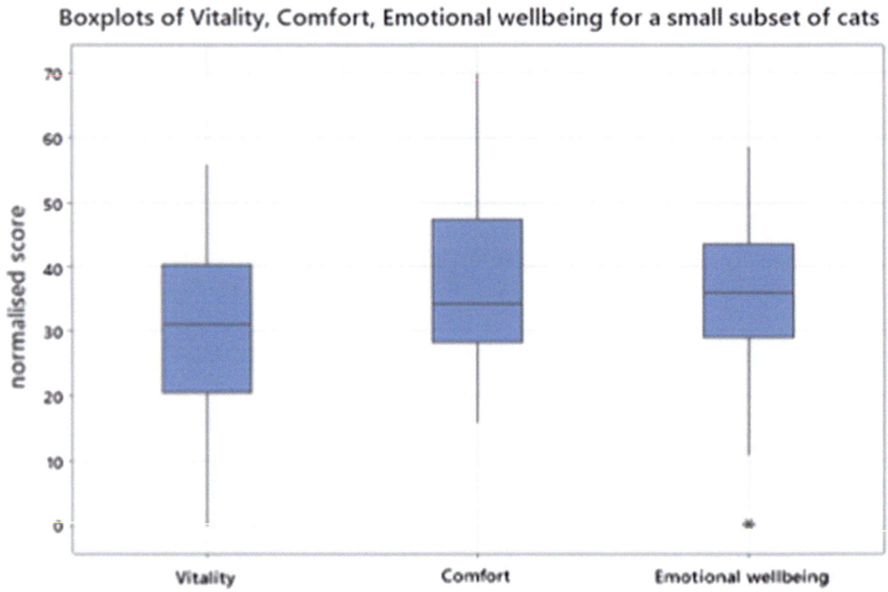

Fig. 1 Boxplots of the domain scores

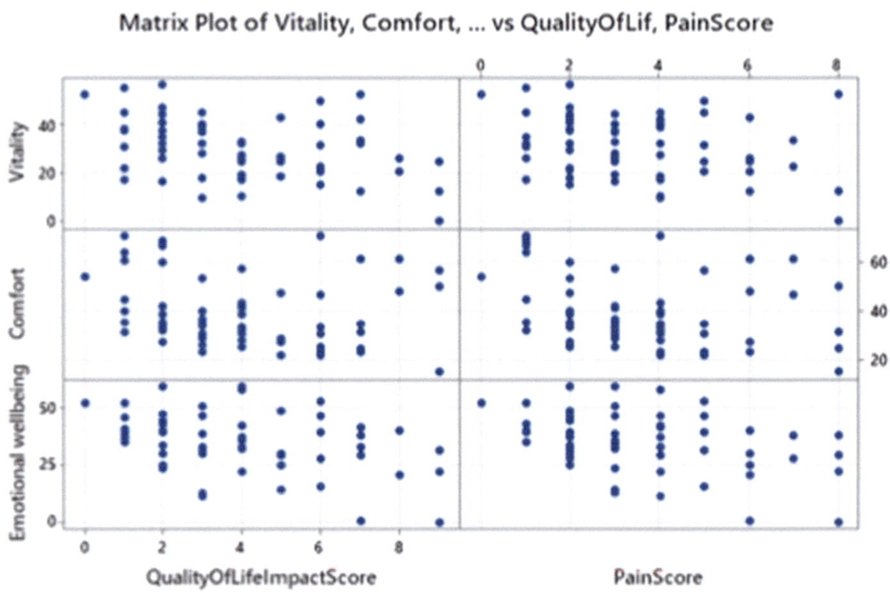

Fig. 2 Scatterplots of the domain scores versus pain and quality of life impact scores

status of the dogs. The performance of the models in predicting the health status was evaluated considering the sensitivity and specificity. A similar development created the alert levels for cats but using the three domains.

4 Impact

Assessment of animal pain and quality of life is critical to veterinary practice. UofG statistical research underpinned the development of robust, psychometrically valid tools to measure acute pain, chronic pain and health related impacts on quality of life in dogs and cats. The SME Newmetrica Ltd. was formed in Nov 2013 to commercialise the University of Glasgow tools, changing veterinary practice in how pain and quality of life is assessed and supporting joint decision-making by owners and veterinarians. Animal healthcare companies also use the tools to support regulatory approvals for new products and markets, and to assess multiple surgical and drug interventions. The Glasgow Composite Measure Pain Scale (CMPS) is considered the gold standard for assessing acute pain in dogs in clinical settings as recommended in World Small Animal Veterinary Association (WSAVA) pain guidelines. Both the acute and chronic assessment tools have been widely used and driven changes including:

- **Change in professional practice within veterinary profession**
  As a result of first paper-based, then web-based and now app-based, pain and quality of life assessment tools, they are being translated into many languages and widely used in veterinary practices, clinics and hospitals. Each translation required cross-cultural validation to define behavioural expressions. Our paper [3] presenting the introduction of facial aspects to clinical pain assessment acute pain in cats, was recognised for its outstanding contributions in the field of small animal veterinary practice at the BSAVA awards in 2016.
  The chronic pain/health related quality of life assessment tool for dogs, having undergone significant practical improvement from its original paper-based version, is now available as a web-based app, and additional statistical modelling has delivered an alert system to inform owners at stages when veterinary advice would be sought [5]. Additionally, a new online tool for cats has been developed, with a new design, in modular form, with a generic QoL module, and bolt on additions for disease specific characteristics.
  Use in clinical veterinary practice
  The development of the first, psychometrically validated acute pain and chronic pain assessment tools for cats and dogs have stimulated and reframed the debate around how pain assessment should be conducted. The acute pain tools are widely used in clinical practice. Online licenses issued between January 2016 and December 2020 for the CMPS-SF tool (dogs) number 2,455 in 76 countries, with the greatest volume in the UK, USA, Spain, Australia and Canada; 90% were specifically for vet practice usage. Over the same period, licenses for the CMPS-Feline

tool number 2,206 in 67 countries, primarily USA, UK, Spain and Canada; 93% were specifically for veterinary clinic use. These numbers reflect changing practice in veterinary clinics, such as in the use of analgesics.

Regulatory changes FDA and medicines approval

In collaboration with Zoetis, supported by a University of Glasgow knowledge exchange award, work was undertaken to ensure that the data resulting from the web-based apps and the subsequent statistical modelling meets the stringent quality assurance needed by the FDA opening the market for wider use in US clinical trials. These tools have provided the pharmaceutical industry with valid, reliable and responsive outcome measures to support their clinical trials for post-marketing and regulatory purposes.

- **Conceptual and cultural impact-assessment of pet health and wellbeing**
  The development of the validated HRQL assessment tools for cats and dogs have stimulated and reframed the debate around how pain assessment and its impact should be conducted. Since the HRQL tools are used by owners in their own home this has raised awareness of subtle changes in pet behaviours by their owners.

Acknowledgements All the owners, vets and veterinary practises who have supported this work.

References

1. Reid, J., et al.: Development of the short-form Glasgow Composite Measure Pain Scale (CMPS-SF) and derivation of an analgesic intervention score. Anim. Welf. **1**, 97–104 (2007)
2. Calvo, G., et al.: Development of a behaviour-based measurement tool with defined intervention level for assessing acute pain in cats. J. Small Anim. Pract. **55**, 622–629 (2014)
3. Holden, E., et al.: Evaluation of facial expression in acute pain in cats. J. Small Anim. Pract. **55**, 615–621 (2014)
4. Wiseman-Orr, M.L. , Scott, E.M. , Reid, J., Nolan, A.M.: Validation of a structured questionnaire as an instrument to measure chronic pain in dogs on the basis of effects on health-related quality of life. Am. J. Vet. Res. **67**, 1826–1836 (2006)
5. Davies, V., Reid, J., Wiseman-Orr, M.L., Scott, E.M.: Optimising outputs from a validated online instrument to measure health-related quality of life (HRQL) in dogs. PLoS ONE **14**, e0221869 (2019)

Mathematical Models Underpin Policy and Practice to Protect Endangered Red Squirrels

Andy White, Hannah Jones, Peter Lurz, and Andrew Slade

Abstract The protection of threatened native species, and control of invasive species, requires the implementation of well-developed management policies. Mathematical models are a key tool that can be used to test, develop and optimise strategies to manage invasive and protect native species. The work outlined in this study represents a long-standing research programme to use mathematical models to inform policy and practice to protect native red squirrels in the UK that are threatened with replacement by the introduced, invasive, grey squirrel. We show how classical deterministic models of interacting species and infectious disease can be extended to spatial, stochastic models that represent the dynamics and habitat of the UK squirrel system. Working with stakeholders, the research has helped shape red squirrel conservation policy and had a direct impact on current and future forest management practice in the UK that balances timber production and red squirrel conservation.

1 Introduction

Since its introduction into the British Isles (in the late 19th century), the North American grey squirrel (*Sciurus carolinensis*) has replaced the native red squirrel (*Sciurus vulgaris*) (Fig. 1) throughout most of England and Wales, and in parts of Scotland and Ireland. As such the red squirrel is a protected species in the UK and was one of the first species identified for conservation under the UK Biodiversity Action Plan.

Mathematical models are key tools that can be used to understand the interactions between red and grey squirrels and to test management strategies that may protect red squirrels. In response to field studies, that indicated that grey squirrels had a competitive advantage over red squirrels in deciduous and mixed woodland, ordinary differential equation (ODE) model frameworks were developed that represented the

A. White (✉) · H. Jones · P. Lurz · A. Slade
Department of Mathematics, Maxwell Institute for Mathematical Sciences, Heriot-Watt University, Edinburgh EH14 4AS, UK
e-mail: a.r.white@hw.ac.uk

P. J. Aston (ed.), *More UK Success Stories in Industrial Mathematics*,
Mathematics in Industry 42, https://doi.org/10.1007/978-3-031-48683-8_24

Fig. 1 A red squirrel (*Sciurus vulgaris*). Photograph courtesy of Peter Lurz

competitive interactions between red and grey squirrels. However, the model results could not match the rate of replacement of red squirrels observed in England and Wales. In addition to competition, red squirrels are also susceptible to a lethal infection, caused by squirrelpox virus (SQPV). The virus, which is believed to have been introduced into the British Isles by grey squirrels, is endemic but harmless to grey squirrel populations. However, SQPV can be transmitted between squirrel species and the infection is lethal to red squirrels, reducing local population abundance by up to 80%. The ODE frameworks that represented the competitive interactions between red and grey squirrels were modified to include the role of SQPV. The model results provided evidence that SQPV infection was a key driver of the rapid replacement of red squirrels by greys in the UK [1] and when combined with empirical data led to the conclusion that SQPV is unequivocally linked to replacement of red squirrels.

The deterministic, ODE, framework allowed the key population processes in the squirrel system to be defined and understood. However, deterministic models do not include the randomness and variability that is exhibited by real systems. Furthermore, to use the mathematical models to inform key policy and management decisions to protect red squirrels required that the models could represent the habitat composition and distribution of the real system, and include the chance of stochastic population extinction and infection fade-out at the local level. The research group, led by Prof White at the Maxwell Institute for Mathematical Sciences, Heriot-Watt University, has developed spatial, stochastic models to represent the red and grey squirrel system and SQPV in the UK and to address key conservation management issues [2–5].

2 Mathematical Model Development

The spatial, stochastic models of the red and grey squirrel system and SQPV in the UK [2–5] are based on the ODE model of Tompkins et al. [1]. The ODE model represents the density of red squirrels that are susceptible (S_R) or infected (I_R) with SQPV and of grey squirrels that are susceptible (S_G), infected (I_G) or that have recovered (R_G) from SQPV. The model is defined as follows:

$$
\begin{aligned}
\frac{dS_G}{dt} &= (a_G - q_G(H_G + c_R H_R))H_G - bS_G - \beta S_G(I_G + I_R) \\
\frac{dI_G}{dt} &= \beta S_G(I_G + I_R) - bI_G - \gamma I_G \\
\frac{dR_G}{dt} &= \gamma I_G - bR_G \\
\frac{dS_R}{dt} &= (a_R - q_R(H_R + c_G H_G))H_R - bS_R - \beta S_R(I_G + I_R) \\
\frac{dI_R}{dt} &= \beta S_R(I_G + I_R) - bI_R - \alpha I_R
\end{aligned}
\tag{1}
$$

where $H_R = S_R + I_R$ and $H_G = S_G + I_G + R_G$ represent the total populations for red and grey squirrels respectively. Here a_G and a_R represent the maximum rates of reproduction for grey and red squirrels respectively, with the natural rate of adult mortality b assumed to be the same for both red and grey squirrels. The competitive effect of grey squirrels on red squirrels is denoted by c_G, whilst that of red squirrels on grey squirrels is denoted by c_R. SQPV is transmitted (both within and between each squirrel species) with coefficient β. Infected red squirrels die due to the disease at rate α and infected greys recover at rate γ. The susceptibilities to crowding (q_R, q_G) are set to ensure the carrying capacity at a specific location is representative of the local habitat composition (see below). Best estimate parameter values are set using empirical data [1].

In the stochastic model the variables that represent the squirrel densities become integer values that represent the population number. The stochastic model is a continuous-time Markov chain, with transition rates given in (Table 1). The model runs on an array of 1 by 1 km grid-squares linked by dispersal. The mathematical model uses forest habitat information obtained from digital landcover data to create a map consisting of a 1 km^2 resolution grid with each grid square containing the proportion of land covered by different tree species as well as the proportion of urban environment. The link between squirrel density, habitat and tree seed crops is well evidenced and so the gridded habitat data is combined with estimates of squirrel density in different habitat types. This produces a red and grey squirrel carrying capacity value for each 1 km grid square which signifies the potential population density that the landscape can support.

3 Example: SQPV Spread Across Southern Scotland

Landcover data that describes the habitat composition of southern Scotland was combined with observed data for the presence of red and grey squirrels to determine the initial distribution of red and grey squirrels in southern Scotland. The spatially explicit stochastic model (Table 1) was used to examine the spread of SQPV throughout southern Scotland following its emergence at the Scottish border in 2005. Prior to the modelling work, conservation policy had focussed on control in regions where

Table 1 Stochastic model events that govern the dynamics that occur within each 1 km grid square. The event that occurs at each time step is chosen at random from a uniform distribution. The events are assumed to be exponentially distributed with the time between events given by $dt = -\ln(z)/R$ where z is a uniform random number between 0 and 1 and $R = \sum[terms]$ (the sum of all the terms in square brackets). Disease transmission can occur from infected squirrels within the focal grid square as well as from the 8 neighbouring grid cells due to daily movement within a core range of radius, $\theta = 0.15$ km. The dispersal term is shown for the class S_G only but is similar for all other classes. Dispersal can occur to a different patch up to 2 km from the focal patch. Grey squirrel control is shown for the class S_G only, but is similar for I_G and R_G. Control of grey squirrels occurs in grid squares for which the trap effort per day T_e is positive. Further details of the model framework and the calculation of parameter values can be found in [4]

Event	Population change	Probability of event
Birth of S_G	$S_G \rightarrow S_G + 1$	$[(a_G - q_G(H_G + c_R H_R))H_G]/R$
Natural Death of S_G	$S_G \rightarrow S_G - 1$	$[bS_G]/R$
Infection of Grey	$S_G \rightarrow S_G - 1, I_G \rightarrow I_G + 1$	$\left[\beta S_G \left((I_G + I_R) + \theta \sum_{Adjacent} (I_G + I_R) + \theta^2 \sum_{Corner} (I_G + I_R)\right)\right]/R$
Natural Death of I_G	$I_G \rightarrow I_G - 1$	$[bI_G]/R$
Recovery of Grey	$I_G \rightarrow I_G - 1, R_G \rightarrow R_G + 1$	$[\gamma I_G]/R$
Natural Death of R_G	$R_G \rightarrow R_G - 1$	$[bR_G]/R$
Birth of S_R	$S_R \rightarrow S_R + 1$	$[(a_R - q_R(H_R + c_G H_G))H_R]/R$
Natural Death of S_R	$S_R \rightarrow S_R - 1$	$[bS_R]/R$
Infection of Red	$S_R \rightarrow S_R - 1, I_R \rightarrow I_R + 1$	$\left[\beta S_R \left((I_G + I_R) + \theta \sum_{Adjacent} (I_G + I_R) + \theta^2 \sum_{Corner} (I_G + I_R)\right)\right]/R$
Natural/Diseased Death of Red	$I_R \rightarrow I_R - 1$	$[(b + \alpha)I_R]/R$
Dispersal of S_G	$S_G \rightarrow S_G - 1, S_G^* \rightarrow S_G^* + 1$	$\left[mS_G \left(\frac{(H_G + c_R H_R)^2}{(K_G)^2}\right)\right]/R$
Control of S_G	$S_G \rightarrow S_G - 1$	$[cT_e S_G]/R$

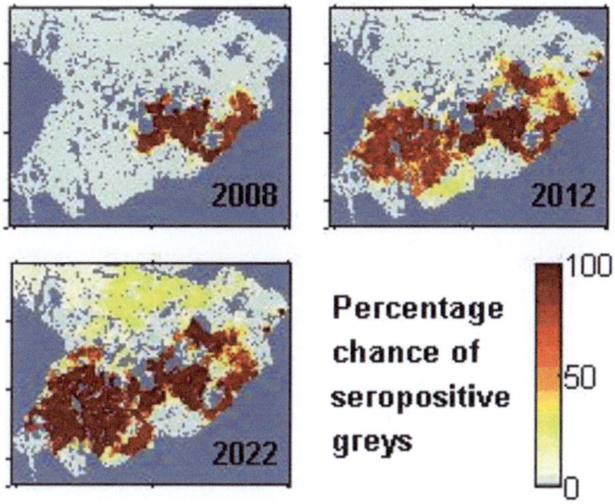

Fig. 2 The percentage occurrence of SQPV over time for a model representing the habitat of southern Scotland. SQPV infection was instigated to represent the observed outbreaks near Canonbie in Dumfries and Galloway in 2005 and near Coldstream and Eyemouth in the Scottish Borders in 2010. The model assumes grey squirrel control is applied in response to SQPV outbreaks and reduces grey squirrel density in control regions. For further details see [2]

grey squirrels tested positive for SQPV. This had reduced grey numbers in local areas but had not prevented disease spread or increased red squirrel numbers. The mathematical model was used to assess the impact of grey squirrel control on preventing the further range expansion of SQPV [2]. The model results indicated that SQPV can spread rapidly through the well-connected habitats of southern Scotland and that grey squirrel control failed to contain SQPV spread (Fig. 2). The work showed that it is extremely challenging to prevent the spread of the SQPV into areas where grey squirrels are already well-established. The modelling work also suggested that SQPV will not persist or spread extensively through red squirrel (only) populations. Saving Scotland's Red Squirrels (SSRS), who are responsible for red squirrel conservation in Scotland, used the modelling work when developing the Scottish Strategy for Red Squirrel Conservation (2015). The policy recognised the difficulty in containing SQPV as the trigger for a change of strategy that protects remaining red squirrel populations in Priority Areas for Red squirrel Conservation (PARCs) in southern Scotland.

4 Impact on Red Squirrel Conservation

The model system outlined in Sect. 3 was extended to answer specific red squirrel conservation issues faced by forest management practitioners. Working with Stakeholders in the Forestry Commission (FC), future felling and restocking scenarios for Kidland and Uswayford forest (a 2962 ha forest in north-east England) were used to drive the mathematical model and led to predictions of red squirrel population viability [3]. The findings highlighted how the (original) proposed forest design plans could lead to a risk of red squirrel extinction (particularly in Uswayford). As a result, alternate felling and restocking regimes were developed by FC and tested in the model. The model results had a direct impact on the forest design plans proposed for the Uswayford and Kidland forest (for 2017–2041) and represent an improved outcome in terms of silviculture and red squirrel viability.

The model system was used to improve grey squirrel control strategies by fitting the model to field observations for the removal of grey squirrels and reintroduction of red squirrels on the Isle of Anglesey, Wales [4]. The model showed that the most effective strategy to remove grey squirrels would be to focus the initial trapping effort on high density regions and then spread the trapping effort to all regions where greys are present. The model also showed that the best strategy to prevent grey squirrel re-invasion of Anglesey would be to focus trapping efforts on the mainland side of the Britannia bridge and use high levels of targeted trapping on Anglesey when grey squirrel sightings are reported. The research findings were adopted by the Red Squirrel Survival Trust (RSST) in 2017 as the strategy to protect red squirrels on Anglesey, where red squirrels are still thriving. The modelling work also played a key role in providing evidence-based research to convince the public of the ongoing need to control grey squirrels.

A collaboration with Forestry and Land Scotland used the model to determine the viability of 19 designated red squirrel strongholds in Scotland [5]. The model compared red squirrel viability under stronghold forest management (SHM) plans (where tree species composition is altered to discourage grey squirrels) and UK forest standard (UKFS) management plans (which allows greater tree species diversity). The research showed that some designated strongholds were unsuitable for red squirrel protection, that other regions may provide natural strongholds (where reds are protected under UKFS plans) and that SHM plans provided no benefit to red squirrel abundance, compared to UKFS, north of the Highland Red Squirrel Protection Line in Scotland [5]. Scottish Forestry intend to undertake a review of SHM policy and the mathematical modelling results will form part of this review. The research findings will allow a future focus on natural strongholds under the UKFS approach and this will afford more flexibility to conserve red squirrel populations whilst simultaneously delivering other forest management objectives.

The model system continues to be adapted to test strategies to protect red squirrels in the UK and to help shape management and policy decisions. This includes a focus on expanding the modelling framework to include ecosystem interactions, such as the impact of predation on grey squirrels from recovering pine marten populations in the British Isles.

Acknowledgements The mathematical modelling work has been supported by Forestry and Land Scotland, Forestry Commission England, the Heritage Lottery Fund, the Natural Environmental Research Council, Nature Scotland, the Red Squirrel Trust Wales and the Scottish Wildlife Trust. HJ was funded by and EPSRC DTA award. AS was supported by the Maxwell Institute Graduate School in Analysis and its Applications, a Centre for Doctoral Training funded by the UK Engineering and Physical Sciences Research Council, the Scottish Funding Council, Heriot-Watt University and the University of Edinburgh.

References

1. Tompkins, D. M., White, A. R. & Boots, M. 2003. Ecological replacement of native red squirrels by invasive greys driven by disease. Ecology Letters. 6: 189–196. (https://doi.org/10.1046/j.1461-0248.2003.00417.x)
2. White, A., Lurz, P. W. W., Bryce, J., Tonkin, M., Ramoo, K., Bamforth, L., Jarrott, A. & Boots, M. 2016. Modelling disease spread in real landscapes: Squirrelpox spread in Southern Scotland as a case study. Hystrix, 27: 1. (https://doi.org/10.4404/hystrix-27.1-11657)
3. Jones, H. E. M., White, A., Geddes, N., Clavey, P., Farries, J., Dearnley, T., Boots, M. & Lurz, P. W. W. 2016. Modelling the impact of forest design plans on an endangered mammal species: the Eurasian red squirrel. Hystrix, 27: 1. (https://doi.org/10.4404/hystrix-27.1-11673)
4. Jones, H.E.M., White, A., Lurz, P.W.W. & Shuttleworth, C.M. 2017. Mathematical models for invasive species management: Grey squirrel control on Anglesey. Ecological Modelling. 359: 276–284. (https://doi.org/10.1016/j.ecolmodel.2017.05.020)
5. Slade, A., White, A., Kortland, K. & Lurz, P.W.W. 2020. An assessment of long-term forest management policy options for red squirrel conservation in Scotland. Hystrix, 31(2). (https://doi.org/10.4404/hystrix-00351-2020)

Controlling the Spread of Livestock Diseases with the Help of Stochastic Models

Max Lau, Simon Firestone, George Streftaris, Glenn Marion, Amy Burroughs, and Gavin Gibson

Abstract Bayesian tools for fitting stochastic epidemic models to complex data sets describing genetic information on pathogens and locations of infections were developed by extending the established approach of data-augmented Markov Chain Monte Carlo. These new methods were adopted by scientists engaged in controlling diseases of farm livestock who extended them by including farm-level covariates and contact information. A user-friendly computer package BORIS (Bayesian Outbreak Reconstruction Inference and Simulation) was developed and applied in the control of real-world epidemics. BORIS's capacity to reconstruct sources (and timings) of infections and to accommodate unobserved infections has proved particularly valuable. Since July 2018, BORIS has improved understanding of the epidemic within the New Zealand Ministry for Primary Industries (MPI) eradication programme for *Mycoplasma bovis*, a bacterial disease affecting cattle, estimated to cost NZD

M. Lau
Rollins School of Public Health, Emory University, 1518 Clifton Rd, Atlanta, GA 1518002, USA
e-mail: msy.lau@emory.edu

S. Firestone
Melbourne Veterinary School, Faculty of Science, University of Melbourne, Parkville 3010, Australia
e-mail: simon.firestone@unimelb.edu.au

G. Streftaris · G. Gibson (✉)
Maxwell Institute for Mathematical Sciences, Department of Actuarial Mathematics and Statistics, Heriot-Watt University, Edinburgh EH14 4AS, UK
e-mail: g.j.gibson@hw.ac.uk

G. Streftaris
e-mail: g.streftaris@hw.ac.uk

G. Marion
Biomathematics and Statistics Scotland, JCMB, The King's Buildings, Peter Guthrie Tait Road, Edinburgh EH9 3FD, UK
e-mail: glenn.marion@bioss.ac.uk

A. Burroughs
Ministry for Primary Industries - Manatū Ahu Matua, Pastoral House, 25 The Terrace, PO Box 2526, Wellington 6140, New Zealand
e-mail: Amy.Burroughs@mpi.govt.nz

P. J. Aston (ed.), *More UK Success Stories in Industrial Mathematics*,
Mathematics in Industry 42, https://doi.org/10.1007/978-3-031-48683-8_25

886,000,000. Specifically, BORIS has been used to infer times and sources for infections to build confidence in the approach to surveillance and control and potentially to identify gaps in surveillance. This eradication programme has been effective with only 5 premises out of more than 60,000 having active disease in November 2022. BORIS has also enhanced understanding of Foot and Mouth Disease (FMD) in Japan, demonstrating its high transmissibility from farms holding predominantly pigs.

1 Introduction

For several decades, stochastic models—which explicitly represent the uncertain nature of events such as infections during an epidemic—have been important tools for informing control strategies for diseases of humans, wildlife, livestock and plants. Many examples can be found in the literature. Typically, statistical inference, whereby data on outbreaks are used to inform estimates of key epidemic parameters, plays a key role.

Why are stochastic models valuable? Firstly, if we can build a stochastic model of an ongoing epidemic with realistic estimates of important parameters, we may predict the distribution of its future course under a range of control measures, to identify those that can most surely mitigate spread with the minimum loss to crops or livestock. The design of the radius of culling (whereby the susceptible population within some radius from a newly discovered infection is removed) for diseases such as Foot and Mouth Disease (FMD), or citrus canker is one such example. Moreover stochastic models can provide valuable insights into unobserved events that have already occurred during an epidemic, to identify potential mechanisms of spread. Suppose we observe an epidemic spreading through a network of farms, but data only give partial information—such as the time of emergence of symptoms on infected farms. If we can identify the precise time of infection of a given farm, and the farm from which infection was acquired, then we can scrutinise interactions between the two farms during the relevant period to help with identification of practices or processes that facilitate the spread of the disease.

In the case of an epidemic spreading locally in space, then simply observing the locations of infections, along with approximate infection times, can provide information on the unobserved infection graph. How might such insights be enhanced by the availability of genetic information on pathogen strains from infected farms? Intuitively, reconstructing the phylogenetic tree for the population of strains detected during an epidemic could provide much information on the transmission graph—given that different locations with genetically similar strains plausibly are more likely to be linked by a transmission event. Therefore, epidemic modellers have devoted much effort to addressing the challenge of integrating genetic information with other forms of epidemic data when fitting epidemic models. We describe one approach that uses ideas from computational Bayesian statistics to develop flexible analytic tools that can support decisions relating to the control of epidemics.

2 Mathematical Model Development

We utilise a standard spatio-temporal Susceptible-Exposed-Infectious-Removed (SEIR) structure. We assume that the locations of hosts, where a 'host' here corresponds to an individual farm, are known. For convenience, we label hosts by their spatial coordinates. The population at time t is partitioned into subsets $S(t)$, $E(t)$, $I(t)$ and $R(t)$. Hosts in $S(.)$ are susceptible to infection, hosts in $E(.)$ have been exposed and infected but are not yet able to infect other hosts, hosts in $I(.)$ can pass on infection, while hosts in $R(.)$ have been removed (e.g. by culling, death or the acquisition of immunity). A susceptible host located at $\mathbf{z}_1$ becomes exposed during $[t, t + dt]$ with probability $C(\mathbf{z}_1, t)\, dt + o(dt)$ where,

$$C(\mathbf{z}_1, t) = \alpha + \beta \sum_{\mathbf{z}_2 \in I(t)} K(\kappa, \mathbf{z}_1, \mathbf{z}_2) \tag{1}$$

Here α and β are *primary* (infection from environmental/external sources) and *secondary* (host-to-host) infection rates, and κ parameterises the *spatial kernel function* $K()$ which represents the way in which infectious challenge between an infected/susceptible pair is related to their relative positions. When models of this form are used to inform 'ring-culling' strategies, then estimation of parameters in the spatial kernel function is of key importance. The SEIR model is completed by specifying distributions, typically Gamma or Weibull, for sojourn times in E and I. We denote by θ the vector of model parameters formed from α, β, κ supplemented by parameters specifying the distributions of sojourn times in E and I.

The main challenge in estimating θ arises from the imperfect observation of real-world epidemics. Available data y may only comprise, for example, times of removals of hosts or 'snapshots' of the infected host population at infrequent sampling times. In such cases, the likelihood function $\pi(y|\theta)$ is typically intractable and investigation of the Bayesian posterior $\pi(\theta|y) \propto \pi(\theta)\pi(y|\theta)$, where $\pi(\theta)$ denotes the parameter prior, is problematic. Data-augmented Markov Chain Monte Carlo (MCMC) overcomes these difficulties by exploiting the fact that the 'full-data' likelihood $\pi(x|\theta)$, where x includes the precise times and nature of all events during the epidemic to date, is tractable. This enables investigation of the joint posterior density $\pi(\theta, x|y)$ using MCMC methods. Standard approaches, including Gibbs sampling and Metropolis-Hastings methods, can be used to update the unobserved transitions in x and explore their posterior distribution. This approach, details of which are given in [1], has been successfully exploited in numerous studies over the past 25 years.

In [1] this framework was extended to include genetic evolution of the pathogen (as characterised by genetic sequences for specified sites on the genome) within infected hosts with the following important assumptions.

(1) The genetic evolution of the pathogen does not affect its transmission characteristics, at least over the time period that the epidemic is observed.
(2) At any time within an infected host, there is a single dominant strain within the host—which is transmitted to any new infection caused by that host at that time. Within-host diversity of the pathogen is not represented in the model.

Assumption (1) means that the model of [1] can be considered to generate the epidemic events (infection times, sources of infection and transition times) and the genetic events (mutations in pathogen sequences) as follows. The epidemic component of the model first generates a framework (or graphical structure) on which the genetic events will occur. This framework is defined by the hosts that become infected, the period during which the pathogen is present in the respective infected host, and the sources of infection for each host. Genetic evolution proceeds on this framework according to a 2-parameter Kimura model for within-host evolution of the single dominant strain. When a new infection occurs, the dominant strain within the infecting host (Assumption (2)) provides the initial state for an independent evolutionary process for the dominant strain within the newly infected host. When the source of infection is environmental (i.e. not an already infected host) the infecting strain is generated as a random mutation of some consensus sequence. These ideas are illustrated in Fig. 1. Assumption (1) facilitates calculation of a complete-data

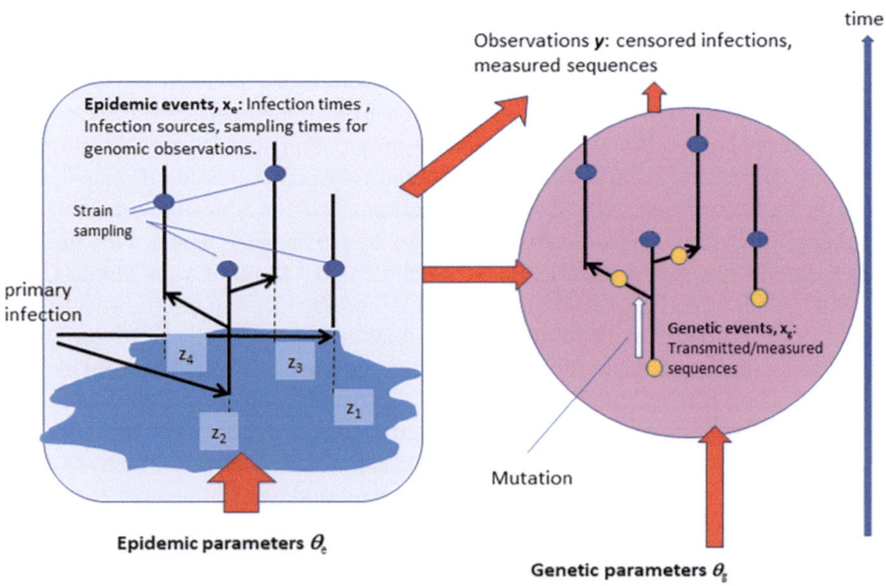

Fig. 1 Schematic diagram of joint model for spatio-temporal epidemic and genetic dynamics. This hypothetical scenario involves 4 farms located at $\mathbf{z}_1, \ldots, \mathbf{z}_4$. Farms $\mathbf{z}_2$ and $\mathbf{z}_3$ are infected via primary infection; $\mathbf{z}_2$ infects the other two farms. Solid lines above each farm represent the period during which pathogen is present and evolving. When combined with sources of infection, these determine the lattice on which genetic evolution occurs.

likelihood for this joint model of epidemic dynamics and pathogen evolution. This has the form $\pi(x_e, x_g|\theta_e, \theta_g) \propto \pi(x_e|\theta_e)\pi(x_g|\theta_g, x_g)$, where θ_e and θ_g are parameter vectors for epidemic and genetic processes respectively and x_e and x_g denote the complete data on these processes. Given partial information on infected hosts and their infection times along with pathogen sequences for (at least some) infected hosts, parameter inference for this model in [1] is achieved by data-augmented Markov Chain Monte Carlo, whereby all unobserved transition times, sources of infection, and transmitted strains are treated as unobserved parameters. Using partial data simulated from the model, the authors of [1] demonstrate the feasibility of the computational approach and explore how the information gained on transmission graphs is related to the proportion of infected hosts for which genomic information is available. Using data from a cluster of farms infected with Foot and Mouth Disease (FMD) during the UK 2001 epidemic, they contrast the transmission graph estimated using their method with earlier estimates using alternative approaches.

3 Benchmarking Performance and the Development of BORIS

The potential of the methods of [1], in comparison to other techniques for integrating epidemic and genomic data, was demonstrated in the study [2]. This study, by a group at the forefront of modelling and controlling ongoing disease outbreaks in livestock, compared the power and robustness of several available methods for reconstructing transmission trees, and other parameters—including molecular-clock rates, using hypothetical epidemics of FMD simulated using complex, highly-parameterised, process-based epidemic models. Of the methods considered, the method of [1] was found to the best for inferring infection sources from partial observation of an epidemic, while performing less well with regard to inference of mutation rates—arguably as a result of the assumption of a single dominant strain. Techniques using more sophisticated genetic models, such as those based on the use of the structured coalescent were found to be perform better in this latter respect.

Given the practical importance of transmission graph estimation for epidemic control, the authors of [2] obtained funding to implement the method as a user-friendly computer package that could be readily deployed to analyse data on emerging epidemics in livestock in real time in order to support decision making in control programmes. The resulting package, named BORIS (Bayesian Outbreak Reconstruction Inference and Simulation), incorporated the algorithm of [1] as the main computational engine along with the additional capacity to include farm-specific covariates (for example representing the species mix on a farm) and additional information on contacts between farms. The main advantages of BORIS include its capacity to fit epidemic dynamics and pathogen phylodynamics within an integrated framework, to represent multiple introductions, and to accommodate partial sampling scenarios, enabling the state of unsampled nodes in a network to be inferred.

4 Application of BORIS in the Control of *M Bovis*

Mycoplasma bovis is a bacterial disease that affects dairy and beef cattle, and a recognised cause of mastitis and respiratory disease with associated impacts on production. It was considered a disease of sufficient significance to the New Zealand farming industry to warrant eradication—dairy representing the second largest export market for the country. The New Zealand epidemic was first detected in 2017 following which a major eradication programme was announced in May 2018. The total cost of the programme was estimated to be NZD 886,000,000 (05-2018) [around £470,000,000], including NZD16,000,000 in lost production and NZD 870,000,000 of response costs needed to fight the cattle disease over a 10-year period [4]. According to [4], had the epidemic been allowed to proceed unchecked the cost would have been approximately NZD 1,300,000,000 (05-2018) [around £689,000,000] over 10 years, with ongoing productivity losses across the New Zealand farming sector.

The BORIS package, code for which is now freely available from github [3], has subsequently been applied within the *M bovis* eradication programme since July 2018 as one of a suite of analytic tools, to analyse epidemiological and genomic data in order to provide key information to support the Programme's understanding of the timing of likely introduction of *M bovis* into New Zealand, whether there were multiple introductions, and the extent to which the case network could be determined. Specifically, BORIS is used to infer timing of infections, infectious periods for farms, and likely transmission routes (who-infected-whom). This serves both to reinforce understanding of who infected whom where the Programme already has a good level of confidence and to explore the uncertainty in the transmission network to assess whether surveillance is providing adequate coverage of possible scenarios, or whether targeted surveillance may be required. Unlike competing genomic methods, BORIS has the capacity to impute chains of infection that include farms for which genetic data on the pathogen may not be available, and to incorporate animal movement patterns during this outbreak as a critical driver of spread between farms. Since the eradication programme was initiated, the epidemic has been successfully controlled to the extent where, as of November 2022 (resp. November 2020), there were only 5 (resp. 6) properties (out of more than 60,000 dairy and beef farms in New Zealand) where the disease was known to be active. BORIS continues to be used as a decision-support tool for guiding surveillance teams within the programme as it moves towards eradication of the disease.

More widely, BORIS been promoted through a training programme on its use delivered by Firestone to veterinary officers at the Australian Government Department of Agriculture, Forestry and Fisheries (Canberra, November 2019). To date 20 scientists have received training in its use. BORIS has also helped improve the understanding of spread of FMD in Japan [5], for example demonstrating the high transmissibility of the disease from farms holding predominantly pigs, and is presently being adapted for use in investigation of transmission processes in the COVID-19 outbreak in Australia.

Acknowledgements Max Lau's research was supported by a scholarship from Biomathematics and Statistics Scotland. Simon Firestone's research is supported by an Australian Research Council Discovery Program grant (DP210103239), the New Zealand Ministry for Primary Industries and the National Institute of Animal Health ((Japan). Glenn Marion is supported by the Scottish Government's Rural and Environment Science and Analytical Services Division (RESAS).

References

1. Lau, M.S.Y., Marion, G., Streftaris, G., Gibson, G (2015) A Systematic Bayesian Integration of Epidemiological and Genetic Data. *PLoS Comput Biol* 11(11): e1004633. https://doi.org/10.1371/journal.pcbi.1004633
2. Firestone, S. M., Hayama, Y., Bradhurst, R., Stevenson, M. A. Reconstructing foot-and-mouth disease outbreaks: a methods comparison of transmission network models. *Nature Scientific Reports* 9(1):4809, https://doi.org/10.1038/s41598-019-41103-6.
3. Firestone S. M., Lau M. S. Y., Kodikara S., Demirhan H., Hayama Y., Yamamoto T., et al. (2019) BORIS: R package for Bayesian Outbreak Reconstruction Inference and Simulation. GitHub repository, https://web.archive.org/web/20201118151729/https://github.com/sfires/BORIS
4. Press release, New Zealand Government, 22/7/2020, *M.bovis eradication makes gains three years on from detection*, https://www.beehive.govt.nz/release/mbovis-eradication-makes-gains-three-years-detection.
5. Firestone, S. M., Hayama Y., Lau, M. S. Y., Yamamoto, T., Nishi T., Bradhurst, R. A., et al. (2020) Transmission network reconstruction for foot-and mouth disease outbreaks incorporating farm-level covariates. PLoS ONE 15(7): e0235660. https://doi.org/10.1371/journal.pone.0235660

Business and Finance

Statistical Modelling for Improving Efficiency of Online Advertising

Irina Scherbakova, Andrey Pepelyshev, Yuri Staroselskiy, Anatoly Zhigljavsky, and Roman Guchenko

Abstract Real-time bidding has transformed the digital advertising landscape, allowing companies to buy website advertising space in a matter of milliseconds in the time it takes a webpage to load. Joint research between Cardiff University and Crimtan has employed statistical modelling in conjunction with machine-learning techniques on big data to develop computer algorithms that can select the most appropriate person to which an ad should be shown. These algorithms have been used to identify suitable bidding strategies for that particular advert in order to make the whole process as profitable as possible for businesses. Crimtan's use of the algorithms have enabled them to improve the service that they offer to clients, save money, make significant efficiency gains and attract new business. This has had a knock-on effect with the clients themselves, who have reported an increase in conversion rates as a result of more targeted, accurate and informed advertising. We have also used mixed Poisson processes for modelling repeat-buying behaviour of online customers. To make numerical comparisons, we use real data collected by Crimtan in the process of running several recent ad campaigns.

I. Scherbakova · Y. Staroselskiy · R. Guchenko
Crimtan, 192 Vauxhall Bridge Road, London SW1V 1DX, UK
e-mail: is@crimtan.com

Y. Staroselskiy
e-mail: yuri@crimtan.com

R. Guchenko
e-mail: rguchenko@crimtan.com

A. Pepelyshev · A. Zhigljavsky (✉)
School of Mathematics, Cardiff University, Senghennydd Road, Cathays, Cardiff CF24 4AG, UK
e-mail: ZhigljavskyAA@cardiff.ac.uk

A. Pepelyshev
e-mail: pepelyshevan@cardiff.ac.uk

P. J. Aston (ed.), *More UK Success Stories in Industrial Mathematics*,
Mathematics in Industry 42, https://doi.org/10.1007/978-3-031-48683-8_26

1 Introduction

The work is devoted to developing models of consumer behaviour for improving efficiency of internet advertisement, where fast decisions have to be made on whether to show a given ad to a particular user. Such decisions are based on information extracted from big data sets containing records of previous impressions, clicks and subsequent purchases.

During the last two decades online advertisement grew enormously, evolved into a major multi-billion dollar industry and became the most significant part of the total advertisement market. The market has been driven by the growth of programmatic buying (when decisions are made by computers rather than people), and more specifically the process of real-time bidding, in which companies are given the opportunity to buy advertising space on websites as part of bidding process in a 'virtual auction'. This process occurs in the milliseconds that it takes for a webpage to load and is repeated billions times each day. Demand partners typically collect databases with logs of all previous requests from auctions, impressions, clicks, conversions and users who visited a website which is currently advertised. These logs typically contain an anonimized user id, a browser name, an OS name, a geographical information derived from the IP address and a webpage link where an auction is run. Merging these datasets with third party data sources provides possibilities for contextual, geographical and behavioural targeting.

We consider the problem of online advertisement via auctions holding by independent ad exchanges from the position of a demand partner which wants to optimise the conversion rate. The demand partner has to decide how reasonable is it showing an ad in regard to a request from an auction and then possibly suggest a bid. The process of showing online advertisements through the so-called Real-Time Bidding (RTB) systems occurs billions of times every day (see [3] for a detailed description of an RTB system).

The demand partner has to solve the problem of maximizing either the click through rate (CTR) or the conversion rate by targeting a set of requests under several constraints: (a) budget (total amount of money available for advertising), (b) number of impressions N_{total} (the total amount of ad exposures), and (c) time (any ad campaign is restricted to a certain time period).

In the rest of the paper, we discuss several issues related to (i) development of specialized machine learning (ML) algorithms for online advertising with addressing the problem of assessing the significance of different factors on probabilities of visits to particular sites, clicks and conversions and (ii) adaptation of the mixed Poisson process model for modelling consumer/user behaviour in internet environment.

1.1 Development of Specialized ML Algorithms for Online Advertising and Comparative Studies

In [1–3] we have made a critical analysis of several procedures for online advertisement, provided a unified point of view on these procedures and have had a close look at the so-called 'look-alike' strategies. In [3], we have also studied relative influence of different factors on the conversion rate and hence develop simple procedures which are very computationally light but achieve the same accuracy as computationally demanding algorithms like Gradient Boosting Machines (GBM) or Field-Aware Factorization Machines (FFM). Note that the number of parameters in the simplest FFM models is the sum of all factor levels plus perhaps interactions between factor levels. It counts to millions and if at least some interactions are taken into account then the count takes to much larger numbers. Unlike standard ML models, the models developed in [3] have a relatively small number of parameters: for predicting the conversion rate, we have propose a sparse model where only a few most significant factors are used.

Formal statement of the problem is as follows. Suppose that the advertisement we want to show is given and first assume that the price for showing a given ad is fixed; we shall also ignore the time constraint. Then the problem can be thought of as an optimization problem for a single optimality criterion which we choose as CTR. We discuss a generic adaptive targeting strategy which should yield the decision whether or not to show the ad to a request from a webpage visited by a user. If the strategy decides to show the ad, it then has to propose a bid. An adaptive decision should depend on the current dataset of impressions and clicks which include all the users to whom we have shown the ad before and those who have clicked on the ad. Note that the dataset size N grows with time. We can increase the size of the dataset by including all our previous impressions of the same or similar advertisements (perhaps applying some calibration to decrease the influence of past ad campaigns), so that N could be very large.

Denote the ith request by $X_i = (x_{i,1}, \ldots, x_{i,m})$, $i = 1, \ldots, N$, where m is the number of features (factors); these features include the behavioural characteristics of the user, characteristics of the website, time of exposure, the device used (e.g. mobile telephone, tablet, PC, etc.). We also assume that for any two points X and $X' \in \mathbb{X}$, we can define a similarity measure $d(X, X')$ which does not have to satisfy mathematical axioms of the distance function.

If $\mathbb{X}$ is a discrete set with all possible requests $X = (x_1, \ldots, x_m) \in \mathbb{X}$ given on the nominal scale then we can use the Hamming distance or the weighted Hamming distance $d(X, X') = \sum_{j=1}^{m} w_j \delta(x_j, x'_j)$, where the coefficients w_j are positive and proportional to the importance of the jth feature (factor), $j = 1, \ldots, m$. These weight coefficients can be computed adaptively on the basis of the analysis of previous data of similar advertising campaigns.

Alternative ways of defining the similarity measure $d(X, X')$ are a logistic model for p_X (as is done in FFM), or to use sequential splitting of the set $\mathbb{X}$ based on the values of the most important factors of X (as in GBM). For FFM, the distance is defined on the space of parameters of the logistic model but in GBM $d(X, X')$ is small if $d(X, X')$ belongs to the same subset of $\mathbb{X}$ and it is large if the subsets which X and X' belong to have been split at early stages of the sequential splitting algorithm (that is, the values of the most influential features are very different).

To identify how the ith factor affects the conversion rate $p(X)$, we use the concept of mutual information. To find the relative influence of the ith factor in the sense the mutual information based on the Shannon entropy, we consider the statistic defined by $I_i^{(Sh)} = \sum_{k=1}^{L_i} \sum_{s=0}^{1} p_{i,k,s} \log_2 p_{i,k,s}/p_{i,k,*}\, p_{i,*,s}$. To find the relative influence of the ith factor in the sense the mutual information based on the Renyi entropy of order α, we consider $I_i^{(Re,\alpha)} = \log_2 \sum_{k=1}^{L_i} \sum_{s=0}^{1} p_{i,k,s}^{\alpha}/p_{i,k,*}^{\alpha-1}\, p_{i,*,s}^{\alpha-1}$. For estimating the conversion rate $p(X)$, we propose $\hat{p}(X) = \sum_{i=1}^{m} I_i q_{i,k_i}/\sum_{i=1}^{m} I_i$ where k_i is such that $X_i = l_{i,k_i}$ and I_i is a relative influence of the ith factor. If we want to use a sparse predictive model then we can set the values of I_i such that $I_i \leq \epsilon$ to zero, for some small $\epsilon > 0$.

In [3] we provide careful comparison of the ML procedure that is based on the use mutual information for selecting the most important features with the classical ones including GBM and FFM. We have shown that the main advantage of the proposed model is its simplicity and time efficiency while the accuracy of all considered ML procedures are practically identical.

1.2 Mixed Poisson Processes for Modelling Dynamics of Visits to Websites, Clicks and Conversions

Mixed Poisson process is a classical probabilistic model used in many different applied areas, where dynamics of occurrence of certain events is of the main interest; see [4]. In consumer behaviour studies, this model can be interpreted as follows. There is a large number of users/buyers. Each buyer makes events (visits, conversions, purchases) according to a Poisson process with his/her own intensity λ. There is a population (mixing) distribution for λ. The most popular mixing distribution is Gamma distribution in which case the model is called Gamma-Poisson and the number of events during a given time period has the negative binomial distribution (NBD). Furthermore, the Dirichlet distribution can be used for modelling brand selection within a category. Resulting model is called Dirichlet market. The basic ideas of this repeat-buying model for consumer research have been developed by A.S.C. Ehrenberg in his fundamental book "Repeat-buying: Facts, theory and applications". We have adapted this model for studying the internet consumer behaviour. As events, we use occasions of a purchase or click or simply of a visit to a given website. The main features distinguishing internet consumer markets with regular ones are: (a) there is much more data available (the total weekly number of visits to gambling or shopping websites could easily exceed millions), (b) there are strong time-of-the-day and time-of-the-week effects (seasonality) illustrated on Fig. 1, and (c) as cookies are not kept long, we have to deal with a very unfavorable effect of not recognizing a

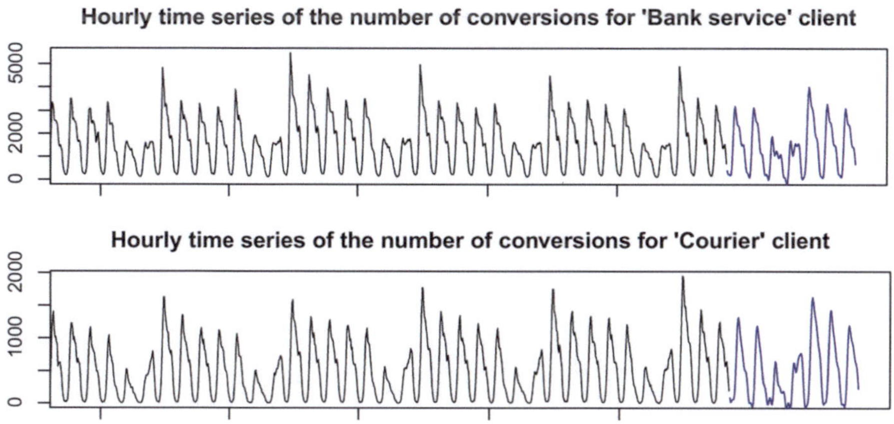

Fig. 1 Typical hourly series for the number of conversions (or simply visits to websites) and their forecasts (blue)

returning user/client after some random time passes; see Figs. 2 and 3 for illustration. Moreover, there is no clear definition of a potential user (buyer) implying that the concept of the total population is undefined. Note, however, that this issue is rather common and not specific for internet markets.

Feature (a) implies that the data we analyse is an instance of "big data" with many features of big data, such as poor quality, present. We routinely use supercomputers for cleansing the data and making various aggregations needed to compute different statistics needed for building the models, parameter estimation, testing the model adequacy and forecasting.

Figure 1 illustrated effect (b). In this figure, we show hourly series for the number of conversions of two different products related to banking service and deliveries in the UK. In both time series, we plot about 6 weeks of data, global peaks are reached on Monday mornings, all global minimima are reached at nights and we can see low numbers during all weekends (they preceed high values of conversions). In view of very clear time trends such time series are easy to forecast with singular spectrum analysis (SSA); see [5]. For the use of the data affected by feature (a) in mixed Poisson process modelling, we need to remove the time heterogeneity by changing the actual time into a virtual time, where each hour contains approximately the same number of events (clicks, visits). In order to do that, we need analytic forms of the forecasts which SSA is conveniently doing for us. These forecasts are also used for early detection of changes in the dynamics of events: an alarm is made when the actual frequencies significantly differ from the forecast values.

Figure 2 illustrates effect (c) and shows the average time of survival of cookies for different types of browsers in days. Estimation of these survival rates are essential for adjusting the mixed Poisson process model. Fortunately, there is large quantities of data and these survival rates can be estimated with high accuracy.

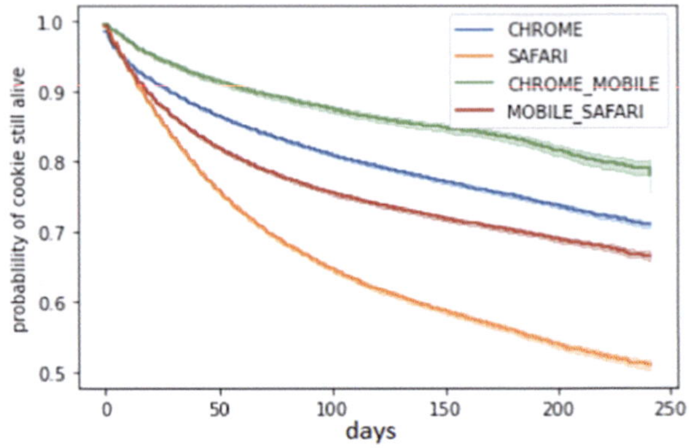

Fig. 2 Cookies survival rates for different types of browsers

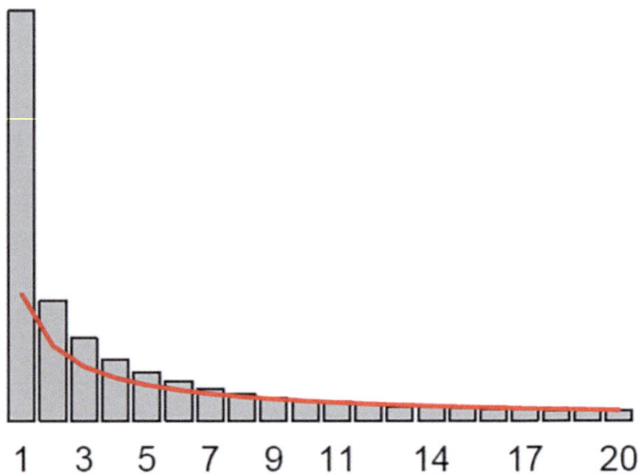

Fig. 3 Frequencies of users visiting a particular website during a given period of time and NBD fit for Gamma-Poisson model in presence of the cookies effect (c)

Figure 3 illustrates that despite the Gamma-Poisson process model can adequately describe the number of visits to a particular website (in this case, a shopping website), the frequencies of visits show larger number of users with small numbers of visits (and hence smaller numbers of loyal users with high numbers of visits), relative to a suitable NBD model. This is caused by our inability of recognizing certain returning users. Depending on the rate of cookies forgetting, we are able to estimate the amount of missing loyal customers and consequently adjust the model.

We use the models based on mixed Poisson processes for the following purposes: (i) studying periodic patterns and trends, (ii) forecasting visits to websites (clicks, sales) for planning marketing activity, (iii) monitoring stability of (early detection of changes in buying behaviours), (iv) studying discrepancies in behaviour of different cohorts of users, (v) modelling loyalty, and (vi) multivariate time series analysis for studying causality and cross-dependence between traffics to different websites and between traffic, clicks and conversions; this includes the use of the Dirichlet market model for modelling brand competition.

2 Impact

Impact has been achieved through the implementation of the computer algorithms in the activites of a London-based digital marketing company, Crimtan (https://crimtan.com/). Crimtan, with offices in 12 countries, delivers dynamic digital marketing campaigns, specifically in the area of real-time bidding, for a large number of clients all over the world. These include Virgin, American Express, Universal, Sky, Volkswagen, Hilton, Visa and HSBC.

With the data at their fingertips, Crimtan have been able to offer their clients highly-targeted advertisement campaigns. The automation of the real-time bidding process has also allowed Crimtan to streamline its own services, resulting in efficiency gains and a significant annual increase in their turnover.

Improved algorithms and using statistical techniques to improve the efficiency and thought processes behind online advertisements have made significant improvements with the accuracy of statistical techniques and hence reduce costs for Crimtan. In particular, the decision-making statistical tools developed in Cardiff helped to reduce the number of human experts employed to make decisions on how and when particular ads can be shown. Experts work efficiently but they are only humans; on the other hand, fast paced computerised calculations nicely emulate the experts and potentially may have a very significant impact.

References

1. Pepelyshev, A., Staroselskiy, Y., Zhigljavsky, A.: Adaptive Targeting for Online Advertisement. In Machine Learning, Optimization, and Big Data. Springer International Publishing (pp. 240-251). (2015)
2. Pepelyshev, A., Staroselskiy, Y., Zhigljavsky, A.: Adaptive Designs for Optimizing Online Advertisement Campaigns. Statistical Papers, vol. 57 (pp. 199-208). (2016)
3. Pepelyshev, A., Staroselskiy, Y., Zhigljavsky, A., Guchenko, R.: Adaptive targeting in online advertisement: models based on relative influence of factors. In: International Workshop on Machine Learning, Optimization, and Big Data (pp. 159-169). 2016. Springer.
4. Grandell, J.: Mixed Poisson processes. Chapman and Hall, 1997.
5. Golyandina N.E, Korobeinikov A., Zhigljavsky A.: Singular spectrum analysis with R. Springer, 2018.

Joining the Dots: Actionable Insights from Interaction Data

Peter Grindrod and Desmond J. Higham

Abstract Published research at the intersection of network science and matrix computations formed the basis of algorithms used by the digital marketing company Bloom. In particular, their product Whisper analysed large scale social media data for commercial insight. Whisper provided results for a range of clients, and formed an essential part of Bloom's offering when Bloom was acquired by Jaywing plc. The challenges raised by the application of network algorithms at scale on real data with stakeholder-defined objectives has opened up further research directions, notably with respect to nonbacktracking walks and higher-order interactions. Here, we introduce the classical Katz centrality measure at the heart of Whisper. We also discuss various recent extensions and emerging applications.

1 Introduction

The digital marketing agency Bloom, based in Leeds, UK, made use of network algorithms in [1, 2] to develop the software product Whisper, which gives insights into large-scale peer-to-peer social interactions. In 2014 Bloom was acquired by Jaywing plc, and Whisper became one of Jaywing's three core products. Here, we highlight some of the subsequent advancements that have been made as a result of (a) our exposure to the challenges that Bloom/Jaywing wished to address and (b) the abundance of new ideas emerging from the network science research community.

P. Grindrod
Mathematical Institute, University of Oxford, Oxford OX2 6GG, UK
e-mail: grindrod@maths.ox.ac.uk

D. J. Higham (✉)
School of Mathematics, University of Edinburgh, Edinburgh EH9 3FD, UK
e-mail: d.j.higham@ed.ac.uk

P. J. Aston (ed.), *More UK Success Stories in Industrial Mathematics*,
Mathematics in Industry 42, https://doi.org/10.1007/978-3-031-48683-8_27

2 Cycling Prohibited

We begin by focusing on directed, unweighted networks (graphs), while noting that a large portion of this material can be extended readily to the case of weighted edges. Letting n denote the number of nodes, a network can be represented via an $n \times n$ binary adjacency matrix A. Here $(A)_{ij} = 1$ if there is a directed edge from node i to node j, and $(A)_{ij} = 0$ otherwise. We also use $i \rightarrow j$ to indicate a directed edge from i to j. We assume that there are no self loops (edges from a node to itself) or multiple edges. We let $D \in \mathbb{R}^{n \times n}$ denote the diagonal matrix such that $(D)_{ii} := (A^2)_{ii}$ counts the number of nodes with which node i shares a reciprocal edge.

The $n \times n$ identity matrix will be denoted by I and we write $\mathbf{1} \in \mathbb{R}^n$ for the vector of all ones. The spectral radius of a matrix C will be denoted $\rho(C)$.

A (directed) *walk of length* r from node i_1 to node i_{r+1} is a sequence of $r + 1$ nodes $i_1, i_2, \ldots, i_{r+1}$ such that the edge $i_\ell \rightarrow i_{\ell+1}$ exists for all $\ell = 1, 2, \ldots, r$. Note that within a walk we may reuse nodes and edges. We can count walks by powering up the matrix A, relying on the fact that $(A^r)_{ij}$ records the number of distinct walks of length r from node i to node j.

Quantifying the importance of nodes in a network is a key task whose relevance is underlined by the success of Google's PageRank algorithm. A *centrality measure* assigns a nonnegative value to each node, with a larger value indicating greater importance. Various centrality measures have been proposed and analysed. Many of these may be motivated by the idea of using traversals around a network to represent the spread of information, influence or disease. Important nodes are those that can efficiently access large portions of the network. Here, walks are both natural and computationally convenient, leading to the use of centrality measures based on matrix power series.

Katz centrality is a widely used matrix function-based centrality measure. Here, we assign a value x_i to each node i by solving the linear algebraic system

$$(I - \alpha A)\,\mathbf{x} = \mathbf{1}, \tag{1}$$

where $\alpha > 0$ is a parameter. Typically A is extremely sparse, so that solving (1) is feasible for very large-scale data sets. The condition $0 < \alpha < 1/\rho(A)$ ensures that $x_i \geq 0$, and we may then expand the resolvent $(I - \alpha A)^{-1} = I + \alpha A + \alpha^2 A^2 + \cdots$, so that x_i may be written as

$$x_i = 1 + \sum_{r=1}^{\infty} \sum_{j=1}^{n} \alpha^r (A^r)_{ij}. \tag{2}$$

We then see that x_i may be viewed as a weighted sum of all walks[1] emanating from node i, with the count for walks of length r scaled by α^r. This scaling

[1] The unit shift on the right hand side of (2) may be regarded as arising from a single walk of length zero. This shift does not affect the relative ordering of the node centralities.

ensures that longer walks carry less weight, which makes intuitive sense from a message-passing perspective. In the limit $\alpha \to 1/\rho(A)$ from below, the Katz centrality vector $\mathbf{x}$ approaches the Perron–Frobenius vector of A, corresponding to eigenvector centrality when the graph is strongly connected.

We note that Katz centrality treats all walks of the same length equally. For example, in the *directed windmill* network in Fig. 1, the walk $7 \to 6 \to 7 \to 6 \to 7 \to 6 \to 7 \to 6 \to 7$ of length eight has the same influence on the centrality measure as $7 \to 1 \to 2 \to 7 \to 3 \to 4 \to 7 \to 5 \to 6$. However, the latter walk traverses the whole network and is hence more likely to be relevant in terms of message passing or disease spreading. This motivates the idea of downweighting the influence of walks that "stutter" around the network. In this vein, the concept of a nonbacktracking walk has proved to be extremely useful—it provides a way to focus on more realistic traversals while maintaining the elegance and computational simplicity of the Katz framework. Moreover, it is reasonable to expect the combinatoric explosion in the number of distinct "stuttering" walks to pose unnecessary constraints on the downweighting parameter α that can be mitigated by eliminating backtracking.

A walk is said to be *backtracking* if it contains at least one subsequence of the form $i \to k \to i$, and *nonbacktracking* otherwise. For example, on the directed windmill above, $3 \to 7 \to 2 \to 7 \to 1$ is backtracking, since it contains $7 \to 2 \to 7$, whereas $3 \to 7 \to 2 \to 1 \to 7$ does not backtrack.

We use $p_r(A)$ to denote the nonbacktracking analogue of the matrix power A^r; that is, the ijth entry of $p_r(A)$ records the number of nonbacktracking walks of length r from node i to node j. A nonbacktracking analogue of Katz centrality (2) is then given by

$$x_i = 1 + \sum_{r=1}^{\infty} \sum_{j=1}^{n} \alpha^r (p_r(A))_{ij}. \tag{3}$$

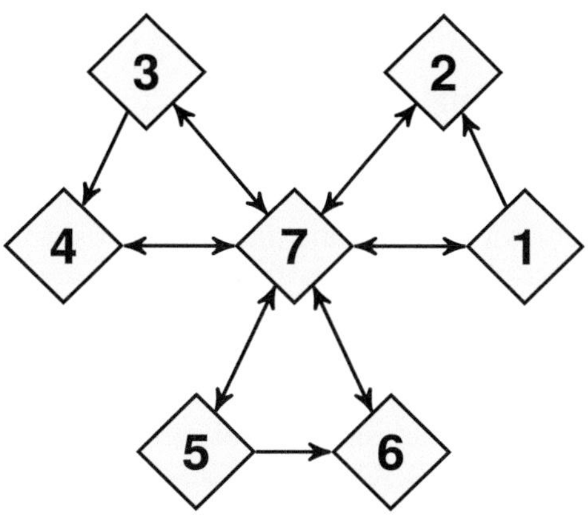

Fig. 1 A directed windmill network

A recurrence expressing $p_k(A)$ in terms of $p_{k-1}(A)$, $p_{k-2}(A)$ and $p_{k-3}(A)$ was originally derived in a number-theoretical setting and has been re-discovered independently in several fields.[2] Using this recurrence, by constructing the appropriate generating function, $\sum_{r=0}^{\infty} \alpha^r p_r(A)$, we showed in [3] that $\mathbf{x}$ in (3) may be computed via the linear system

$$\left(I - \alpha A + \alpha^2(D - I) + \alpha^3(A - S)\right)\mathbf{x} = (1 - \alpha^2)\mathbf{1}. \tag{4}$$

Here $S \in \mathbb{R}^{n\times n}$ has the form $(S)_{ij} = (A)_{ij}(A)_{ji}$ and hence records reciprocated links. Comparing (4) with (1) we see the remarkable fact that *a nonbacktracking version of Katz centrality is available from a linear system that has the same level of sparsity as A*. Hence, nonbacktracking comes for free.

There are three distinct senses in which it may be argued that nonbacktracking improves Katz centrality. First, nonbacktracking can be shown to guard against *localization*; the undesirable circumstance where a network centrality measure focuses almost exclusively on a subset of the network, and does not give useful information about the relative importance of the majority of nodes. Second, moving from (1) to (4) cannot worsen and typically improves the radius of convergence of the associated power series, allowing for a larger downweighting parameter so that *global features of the network have a greater influence*. Third, in the nonbacktracking setting, we may prune the network by recursively removing leaf nodes, source nodes and dangling nodes, thereby reducing the dimension of the problem.

A promising future direction for this active research area, which was opened up in [4], is to *move beyond nonbacktracking and develop new theory and new tools based on noncycling*. The idea of nonbacktracking is to ignore walks with $i \to k \to i$ subsequences, which may be regarded as cycles of length two. In [4] it is shown how to eliminate triangles, $i \to k \to l \to i$, squares, $i \to k \to l \to m \to i$, and, generally, cycles up to length $n - 1$.

3 Think Ahead

The concepts and methods applicable for digital marketing (and targeting) are highly relevant to many other fields. Not least amongst these are *national security* and the protection of citizens from *online harms*. For example, in analysing national and international movements that are socially and politically inspired: extreme political views; anti-establishment anti-authority views; terrorist or insurgent activities; cults; fake news; and so on.

In [5], analysis of social media from (relatively unregulated) peer-to-peer platforms such as Telegram emphasised how various connected digital communities can influence one another, and thus raised the prospect of international, coordinated activities that need to be addressed by various separate national security interests. It was

[2] We have even rediscovered it ourselves on more than one occasion!.

General Stanley McChrystal,[3] the US commander in Afghanistan facing the Taliban and al Qaeda, who said,"*It takes a network to defeat a network.*" Both adversaries were deeply embedded within Afghanistan's complex society, and more like a network than an army, more of a community of interest than a corporate structure. Yet digital platforms enable such groups to operate across regions and national frontiers.

The flows of influence around digital networks result in directed, and often time-dependent, evolving, graphs. The natural asymmetries introduced by successive time-stamps are key. Core-periphery concepts are also important as it is the *circularities of influence* that result in echo chambers within the deep core; these may excite and drive radical thinking.

Such urgent challenges require a firm response from those working in mathematics of networks and social influence. Moreover, the requirements of security and defence exploiters make computational efficiency an essential ingredient. Enabling the analysis of live data and producing "rolling" outputs are key challenges. One of us took this on directly, in part, by analysing a simulated terrorist attack at London Heathrow, which is illustrated by the video https://www.youtube.com/watch?v=hqRrUozG0hY.

We finish by highlighting a further issue within network science, and other data scientific methods, including topological data analysis (persistent homology). Data will contain errors: both false positives and false negatives. The impact of these is not the same, false positives being far more pernicious for modelling activities. Errors may arise due to direct mismeasurement, or else misinterpretation, where we view a range of activities as proxies for communication, influence, reciprocated friendship, and so on; inevitably we make mistakes in such inferences. Those mathematical tools that are sensitive to each and every piece of data are inherently unstable. For this reason spectral methods, relaxation methods, and other techniques that avoid "all-or-nothing" steps, are to be preferred. However, even then the algorithmics may simply be too slow when applied at scale, and hence may be incapable of producing as-live outputs and actionable advice. What is missing is the network theory equivalent of *asymptotic methods*. In many areas of applied mathematics we enjoy allowing some small parameter to tend to zero, so that in a singular limit the problem decouples into separate components that might be solved much more easily: then these solutions are matched (or coerced) together so that we create a combined solution for the whole that is valid in the singular limit. Somewhat mysteriously, we often find that such asymptotic solutions remain valid and useful even when the small parameter is not so small and we are some way from the singular limit. The challenge of creating huge efficiencies in the analysis of vast networks, or real time data avalanches, is a real one that cannot be addressed by the analysis *toy* data sets. The "asymptotics of network science" should give users options to trade efficiencies with accuracies. This is a bottleneck for the present science whenever it is applied to live fields with incoming data hosepipes.

[3] https://foreignpolicy.com/2011/02/21/it-takes-a-network/.

4 Impact

We finish with a quote from the University of Strathclyde REF 2021 Impact Case Study: *Developments in digital marketing through computational analysis of dynamic social media networks*[4]

> Responding to pressure from its competitors, particularly BT Sport who had claimed the rights to air the UK Champions League Football, in 2015 Sky Sports commissioned Bloom to undertake analysis to inform a social media campaign to boost its profile amongst football fans. As outlined in a Jaywing case study, by analysing more than 80 million tweets over a two-month period, Bloom was 'able to visualise how the different football-related communities were connected and who was influential in sharing content and messages between them' [S7]. This insight told them that 'focusing the creative campaign on 'moments' in Premier League history would resonate most with fans', enabling them 'to pinpoint the moments most likely to be shared socially' and provide 'Sky with a list of major influencers and a strategy for how best to engage with them' [S7]. This led to the development of the #PLMoments campaign and creation of an advert which went viral as soon as it was launched on Facebook thanks to the influencers who had been identified through the Whisper analysis. The case study highlights the significance of this for Sky Sports: 'Within the first hour of the video going live it gained over one million views. It also topped the Campaign Viral Chart and has had a total of well over 25 million views. It was the 4th most viewed video of any kind on the internet during its first week and @skysports was the most influential brand on social media during the Premier League weekend' [S7]. Such was the success of this project that it netted Bloom a Social Media Buzz award in 2015 and was highlighted in reports of Jaywing's acquisition of Bloom, with one article explaining that 'Whisper is the intelligence behind Sky Sport's Premier League advert featuring Thierry Henry and iconic moments from Premier League football' [S8].

Acknowledgements DJH was supported by the Engineering and Physical Sciences Research Council under grants EP/P020720/1 and EP/V015605/1.

[4] Citations in this quote refer to the bibliography in the case study.

References

1. Peter Grindrod and Desmond J. Higham. A dynamical systems view of network centrality. *Proc. R. Soc. A*, 470:20130835.
2. P. Grindrod and D. J. Higham. A matrix iteration for dynamic network summaries. *SIAM Review*, 55:118–128, 2013.
3. Francesca Arrigo, Peter Grindrod, Desmond J. Higham, and Vanni Noferini. Non-backtracking walk centrality for directed networks. *Journal of Complex Networks*, 6(1):54–78, 2018.
4. Francesca Arrigo, Vanni Noferini, and Desmond J. Higham. Beyond non-backtracking: non-cycling network centrality measures. *Proceedings of the Royal Society, Series A*, 476, 2020.
5. A. Bovet and P. Grindrod. Organization and evolution of the uk far-right on Telegram. https://assets.researchsquare.com/files/rs-1964336/v1_covered.pdf?c=1660843265, 2022.

Digital Banking—Building Atom bank's Digital Twin

Camila C. S. Caiado, Nathan Huntley, and Edward Twiddy

Abstract Durham University worked in collaboration with Atom bank to develop an end-to-end banking model based on mathematical and statistical methods developed at Durham. The model provides the business with a strategic understanding of the relationships between the bank's key components including customers and products and supports major business decisions such as financial planning, resourcing, product pricing and funding. The impacts of this on Atom bank include improved management of short term liquidity requirements, improved capital deployment and operational risk reduction. Direct savings to the bank are estimated at over £1 million annually since 2018, providing the business with both a competitive edge and an opportunity to share efficiencies with its customers.

1 Underpinning Research: Bayes Linear Methods and Complex Systems Modelling

Mathematical models, implemented as high-dimensional computer simulators, are often used to study complex systems. In general terms, a bank is a complex system with multiple interacting components driven by statistical and financial models that are used to predict future market behaviour and optimise scenarios. Banking models need to be constructed so that they can interact with each other seamlessly and so that they can be easily used to predict behaviours (financial market, customer interactions), optimise strategies (investments, funding capture, pricing) and manage risk

C. C. S. Caiado (✉)
Department of Mathematical Sciences, Durham University, Upper Mountjoy Campus, Stockton Road, Durham DH1 3LE, UK
e-mail: c.c.d.s.caiado@durham.ac.uk

N. Huntley · E. Twiddy
Atom Bank, Rivergreen Centre, Aykley Heads, Durham DH1 5TS, UK
e-mail: nathan.huntley@atombank.co.uk

E. Twiddy
e-mail: edward.twiddy@atombank.co.uk

P. J. Aston (ed.), *More UK Success Stories in Industrial Mathematics*,
Mathematics in Industry 42, https://doi.org/10.1007/978-3-031-48683-8_28

(operational, risk appetite). Based on the Bayes Linear methodology and uncertainty quantification framework [2, 3], Atom bank and Durham University developed a number of critical components necessary for financial operations and decision support solutions used by stakeholders across the business to understand the impact of their choices on the overall business leading to a new interactive and innovative approach to banking.

Durham research on complex simulations, statistical modelling, and optimisation [2–4] was used to mathematically model and optimise the relationship between investments, customer behaviour, market behaviour, and their constraints and impacts. It resulted in a "simulator" of the bank, or the "Atom Digital Twin". Using advanced uncertainty analysis, a statistical model was constructed including an optimisation tool and user interfaces (see Fig. 1) for decision support in line with the work in [2], for example.

In particular, a Bayes Linear approach [2] was used to construct emulators of some of the sub-models of the financial and operational master models of the bank, and history matching was used to identify areas of the models' input spaces (formed by a number of variables the bank can normally control such as prices and staffing).

One of the tools delivered is a funding optimiser that identifies the potential price combinations that can be set in order to minimise future cost; for example, if there were 4 fixed-term mortgage products with different maturity lengths that have their prices updated weekly within certain constrained price ranges. As in [2, 3], we started by building a Bayes Linear emulator of one of the mortgage pricing models.

The objective is to find combinations of prices that minimise costs (and maximise long-term profit where possible) over a period of 5 years; with prices varying weekly

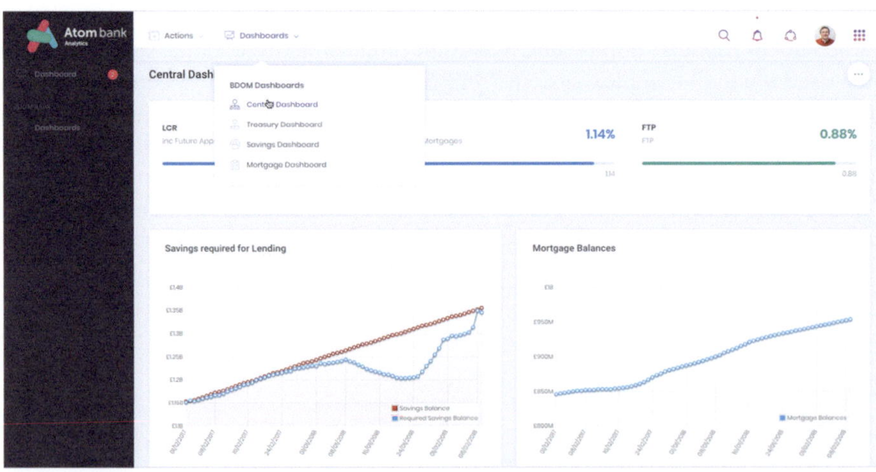

Fig. 1 One of the dashboards produced for stakeholders to interact with the models developed

for each product. In this context, it is necessary to identify suitable time series solutions (length 260 for each product) that are feasible individually but also as a group to ensure capital requirements can be met each week.

Since optimising for 1,040 inputs simultaneously in a reasonable amount of time wasn't feasible, we used history matching and sensitivity analysis to identify the periods in time where price variations were most likely to cause large changes to the output or drive the models into infeasibility. This approach was very effective in reducing the input space and narrowing down price ranges for each product but also to allow the decision makers to see alternative pricing profiles that could lead to a similar (if not the same) outcome that they were used to.

At a given point in time, the bank has the responsibility to retain the correct amount to be paid to customers as their mortgages come to an end, but also has to hold certain amounts of cash to ensure business continuity. However, holding on to money without investing it is costly, so they need to balance out investments and regulatory demands to minimise cost. First, they need to understand how much business to expect if they are to set a specific price for a given product, subject to potential competitor behaviour and other external factors, to derive a price elasticity model of demand for each product. Given a certain demand level, estimated using agent-based modelling principles in [4], a retention profile has to be derived and the minimum cash requirements updated. In [4], the research looked at how customers behave with respect to peer pressure, innovation, product loyalty and memory. For a new bank, loyalty doesn't yet exist and memory needs to be built as part of the retention process. The Durham research looked at the trade-offs between parameters such as for innovation and memory for different customer groups at Atom to estimate retention rates. These values have then been subsequently updated based on real data as products matured.

Even though the individual models mentioned have been optimised to run in real-time (<10 ms), it is necessary to identify the regions in the input space for which cost is minimized. At the moment, this takes around 90s per planning year reducing severely the time taken by this process which used to take hours and required constant manual intervention; a combination of Bayesian optimisation techniques for space reduction, and Bayes Linear methods were used. Now this is enveloped in a user interface that can be used to stress test price interactions, inform treasury of needed funding capture, investigate marketing strategies, and optimise for the yearly 7 year financial planning round.

2 Details of the Impact

Atom bank is a digital challenger bank based in the City of Durham, which started trading in 2016, offering an innovative and efficient mobile-only app-based personal and business banking experience. One of the reasons Atom bank chose Durham as its base was to take advantage of the research output of the University.

> It quickly became apparent that the department's specialisation in applying Bayesian techniques to financial and other systems (e.g. [2, 4]) offered the opportunity to create a digital twin of some of the bank's key allocation decisions.—Edward Twiddy

The collaboration between Atom bank and Durham University resulted in the creation of the Atom bank Digital Twin using methods developed within the Mathematical Sciences department [2–4]. This has allowed Atom bank to re-invent financial planning in financial services enabling it to:

1. Optimise business activity so that it generates the maximum return from each set of investment decisions, and
2. Understand in full the consequential impacts of any business decision on each business area. This enables the bank to link its resource planning accurately to product and pricing decisions. Atom bank says that this was *'critical to building a competitive advantage that can be shared between investors and customers'*. The total estimated savings, detailed below, are currently over £1 million per annum since 2018.

The Atom bank Digital Twin is a novel decision support suite, which is being used to support major business decisions such as financial planning, resourcing, product pricing and funding. Atom bank now has the unique opportunity to mathematically model the organisation and manage the business according to the inputs and outputs of that model. The direct impact of the collaboration with Durham University (which originally centred around a Knowledge Transfer Partnership) on Atom's financial performance can be summarised in three categories.

2.1 Improved Management of Short-Term Liquidity Requirements

Banks are required by regulation to hold sufficient liquid assets to cover their forthcoming cash flows. Many aspects of inflows and outflows, such as the quantity of loans being provided to customers, and the savings deposits received by customers, are uncertain. The models created with Durham research [2–4] have greatly increased the accuracy of Atom's ability to forecast these cash flows and so reduce the overall liquid assets the bank is obliged to hold (which are an overall cost to the business). The main items are:

1. Improved modelling of loan completions allowed the liquidity held against future loan completions to be reduced by approximately 50% at all times, leading to substantial reduction in liquid assets held. This benefit was first fully realised in 2018 and a similar annual benefit remains today.
2. Improved ability to predict the behaviour of savings customers has allowed Atom to reduce the volume of liquid assets needed to ensure that deposit holders can access their funds upon maturity. In 2018 and 2019, it was estimated that a substantial lower amount of liquid assets were needed. In 2020, a different business

strategy meant that the modelling reduced the risk of accidental breaches of regulatory limits, which cannot easily be measured in terms of financial value, but is of critical reputational importance.
3. Advanced uncertainty analysis (R) on the above models allowed the business to reduce buffers (designed to minimise the risk of regulatory breaches) on liquidity, leading to another substantially reduced liquidity requirements in 2019. Again, in 2020, this methodology has led to risk reduction rather than financial improvements.

2.2 *Improved Capital Deployment*

Banks, particularly fast-growing banks such as Atom, are often constrained by available capital. They are obliged to hold certain levels of capital for each loan sold. Therefore, maximising the efficiency of this limited resource is key to profitability. Working with Durham University has principally provided benefits in the following areas:

1. Quality assurance work carried out on Atom's internal lending models is estimated to have led to at least a 0.01% improvement on margin.
2. Improved long-term financial planning from the summer of 2017 onwards will have led to a more efficient business plan. This benefit is significantly harder to estimate than any of the others, but financial planning experts quantified this as at least 0.01% improvement on the net interest margin of the entire book.

2.3 *Operational Risk Reduction*

Another use of a bank's capital is to be held against risks to implementing the business plan successfully. Collaborating with Durham University meant a number of operational risks were reduced. The highlights were:

1. Quality assurance on regulatory mortgage calculations.
2. Reduction in model risk, i.e. the risk that Atom's internal models give erroneous or misleading forecasts.

The financial impact of such risk reductions is notoriously difficult to estimate, however scenario analysis and subjective expert judgements suggest that Atom's operational risk capital allocation may have been substantially higher without collaborating with Durham University. The corresponding reduction in lending would lead to a large reduction in interest income (using the bank's current net interest margin) in 2020 and beyond.

2.4 Changes in Practice at Atom bank

Embedding the use of the Atom bank Digital Twin has had a substantial impact on processes and policy within Atom bank. Atom's data and analytics team has expanded from 4 to 16 members employed in the North East of England to deliver this project and its extensions (gross value added based on North East 2017 figure is £575,112) and a total of 30 staff have been trained in using the model output as well as in the R language to allow model development and maintenance to continue. By embedding expertise in the open-source R programming language, the business has been able to avoid costly licensing for commercial data analysis software. Atom says that the collaboration *'has also shifted expectations and perceptions amongst the senior team at Atom, many of whom had not had or seen the benefit of deep engagement with university researchers in their previous roles'*.

Atom bank have continued to win industry recognition and have recently been included in the Tech Nation Future Fifty business community. They continue to increase the number of products offered and in 2019 total assets growth compared to 2018 was 41.64% [5].

2.5 Impact on Customers of Atom bank

Atom bank is a retail business and so every efficiency improvement also has a benefit to customers. *'Reducing the friction between teams, increasing the allocative efficiency of our capital, and responding to the constraints on optimisation that are a necessary part of being a regulated bank all mean that customers get better prices for their savings and lower costs for their lending than we could otherwise afford to offer. Atom continues to be recognised for its excellence in the savings and lending markets'* (as recognised by various media organisations such as The Times and The Northern Echo) *'and behind all of these sits an engine that is refined and tuned by the work done with Prof Caiado and colleagues.'*. Atom is Trustpilot's most trusted UK bank and consistently achieves Net Promoter Scores in excess of +75.

In summary Durham University has helped to develop an end-to-end banking model that provides a live and interactive overview of the business with a strategic understanding of the relationships between the bank's key components including customers and products. This is unique within the banking sector and the change in practice has given Atom bank the opportunity to update both the simulation and the optimisation in real time as the bank evolves and grows, increasing sustainability and providing rigorous calculations supporting the mortgage lending process that minimise the risk of operational damage or regulatory breaches from calculation error.

3 Current Relationship Between Atom bank and Durham University

Atom bank and Durham University continue to collaborate in a number of areas of research and development related to both Atom's strategic intentions to apply innovative and novel techniques across Atom's operations and support high quality learning and work opportunities and to Durham's aim to achieve impact and develop new research directions. This has been formalized under a memorandum of understanding signed on September 15th, 2022.

The objectives of the partnership are to:

- Carry out research that is collaborative, participative and inter-disciplinary to address longer term opportunities and challenges facing the Parties and society;
- Exchange knowledge to leverage existing know-how, expertise and data to address shorter term opportunities;
- Achieve substantial impact;
- Develop approaches to research that enable it to be put into practice;
- Provide personal development opportunities through industry exposure for early and mid-career university researchers and students and engagement in research with academic staff and students for company staff;
- Discover and develop new, mutually beneficial partnerships founded on both parties' existing partnerships; and
- Support regional economic development and innovative approaches to sustainable business growth.

References

1. Atom bank Annual Report, 2020, https://www.atombank.co.uk/~/docs/annual-report-19-20.pdf
2. Goldstein, M., Wooff, D. (2007). Bayes linear statistics: Theory and methods (Vol. 716). John Wiley & Sons. https://doi.org/10.1002/pst.328
3. Goldstein, M., Huntley, N. (2016). Bayes Linear Emulation, History Matching, and Forecasting for Complex Computer Simulators. Handbook of Uncertainty Quantification, 1–24. https://doi.org/10.1007/978-3-319-11259-6_14-1
4. Bentley, R.A., Caiado, C.C.S. Ormerod, P. (2014). Effects of memory on spatial heterogeneity in neutrally transmitted culture. Evolution and Human Behavior 35(4): 257–263.https://doi.org/10.1016/j.evolhumbehav.2014.02.001
5. The Banks, Total assets growth https://thebanks.eu/banks/18618

Energy

Bringing Cost Savings and Customer Benefits at Electricity North West Ltd

John Moriarty

Abstract Conventional methods of cost-benefit analysis (CBA) under-value investment in 'smart' technologies. To overcome this challenge the Real Options CBA model (RO CBA), a new decision-making process and supporting software, were developed in collaboration with Electricity North West Ltd (ENWL), the electricity distribution network for north-west England. RO CBA adds a stochastic process framework based on Real Options Analysis to a set of long-term forecast scenarios, enabling competing conventional and smart technologies to be compared on a common basis. ENWL has integrated RO CBA into its regular business practice, directly benefitting over 100,000 customers across fourteen different investment projects. This change has enabled the network to provide better-value electricity for its 5 million customers, and has also informed the UK National Grid's best practices on the use of Real Options in cost-benefit analyses.

1 Introduction

Electrical grids around the world are becoming 'smarter'. Through continued innovation and with the increased use of measurement, communications and control, new technologies are allowing electrical power systems to be operated more flexibly, driving increases in both operational efficiency—the cost of running the power system from day to day—and capital efficiency, the level of investment required in the fabric of the system [1].

Increased efficiency is an important justification for investment in such novel flexible technologies. The very flexibility which makes these technologies more efficient, however, also makes their valuation arguably a non-standard task. As a result it can be challenging to construct acceptable business cases for investment in new, smart technologies for the power grid.

J. Moriarty (✉)
School of Mathematical Sciences, Queen Mary University of London, Mile End Road, London E1 4NS, UK
e-mail: j.moriarty@qmul.ac.uk

P. J. Aston (ed.), *More UK Success Stories in Industrial Mathematics*,
Mathematics in Industry 42, https://doi.org/10.1007/978-3-031-48683-8_29

Conventional methods of cost-benefit analysis (CBA) (see, for example, [2]) do not model flexibility—and to the extent that any aspect is not modelled, it is also not assigned economic value. In this way CBA can be said to under-value investment in 'smart' technologies. This makes it highly challenging for networks to compare flexible and inflexible network investments fairly, to justify 'smart' network upgrades, and to obtain best value for money by investing in energy-use innovation. In addition, conventional CBA does not explicitly model real-world uncertainty over future values of the most important variables driving distribution network planning, such as load growth.

Through more bespoke modelling such as stochastic methods, mathematics offers a route through this quandary. Since flexibility has no value in the absence of variation, it is important to consider a range of likely scenarios. This range can be conveniently framed in the language of probability through the introduction of multiple possible outcomes in a sample space. The response of flexible technologies to unpredictable variations can then be modelled through a function defined on the outcome: in probabilistic terms, a random variable. This random variable should appropriately reflect the effect of both the unforeseen variation and the flexible response of the smart technology to that variation. Through consideration of appropriate statistics of these random variables, conclusions can be drawn regarding the balance between the costs and benefits associated with smart grid technologies.

Importantly, both flexible and inflexible technologies can be modelled on a common basis in such a framework. The modelling of inflexible technologies is more straightforward, since the associated random variables need only capture the effect of the random variation upon the inflexible technology, rather than modelling any additional flexible response offered by the technology itself. This common basis is important for business cases which aim to distinguish between a range of possible investments, some of which may be flexible and some inflexible.

In academic research, the stochastic modelling of real-world investments is the domain of Real Options Analysis (ROA). The interdisciplinary collaboration described in the present article, sponsored by Electricity North West Ltd. (ENWL) and involving the University of Manchester and Queen Mary University of London, was undertaken in order to develop a new CBA model based on Real Options Analysis. The project was motivated by the Ofgem consultation document [3], which addressed the potential incorporation of uncertainty in the planning of gas distribution network investments. By appropriately quantifying the economic value of flexibility, its objective was to provide a consistent basis for the valuation of both flexible and inflexible investments. If successfully achieved, this would in turn provide a business process enabling ENWL to identify choices representing better value for money for electricity customers, supporting these choices with the necessary business cases.

2 Smart Grid Technologies, Cost-Benefit and Real Options Analyses

An example of a smart grid technology providing increased flexibility in power networks is demand-side response (DSR) [1]. DSR employs communication and control in order to reduce and redistribute peak-time energy use. This acts to lessen the maximum load, for example at a given point or throughout the system, thus mitigating strain and using resources more efficiently by releasing unused network capacity. The competing conventional alternative is to invest in more expensive infrastructure such as additional or upgraded transformers and overhead lines. Such conventional infrastructure offers the advantage of adding significant, reliable peak capacity to a network location over its lifespan of years and, perhaps, decades.

In the context of investment decision making, however, this long lifespan can have the practical disadvantage of asset stranding. Once a distribution network upgrade is commissioned at a particular location, its associated fixed costs can be largely irrecoverable. Thus if in actuality the level of additional peak capacity needed is lower than anticipated, investment capital can be stranded in an unproductive asset. In contrast, smart grid solutions such as DSR can have largely variable operating costs which scale proportionately to their actual utilisation, such as payments to those end users who reduce or redistribute their own peak-time power use. Any irrecoverable costs associated with smart grid solutions are also typically significantly lower than those of conventional investments, reducing the severity of the problem of stranding.

Figure 1 (reproduced from [4], where further detail can be found) illustrates three scenarios for the evolution of peak demand over a period of 10 years. In scenarios S1 and S2 it is optimal to invest in a conventional asset to ensure that network capacity continues to meet demand over time. In scenario S3, which represents low demand, it is preferable to deploy a limited amount of DSR (here labelled ‘C2C’ for the Electricity North West *Capacity to Customers* DSR scheme) for a limited period of

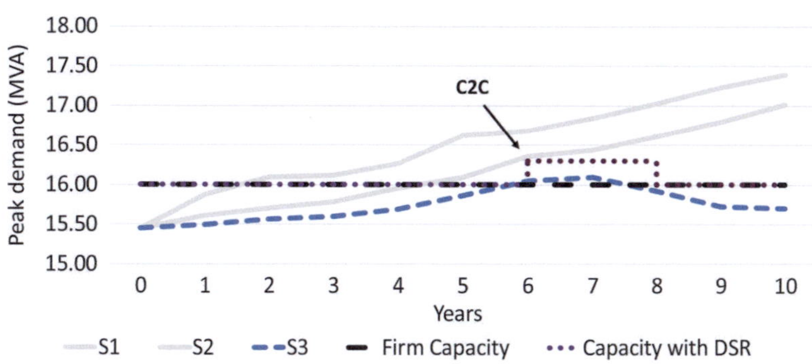

Fig. 1 An example scenario in which it is preferable to deploy a limited amount of DSR (labelled ‘C2C’) for a limited period of two years, rather than making a larger, irrecoverable investment in a conventional asset which would be stranded. (Reproduced from [4])

two years, rather than making a larger, irrecoverable investment in a conventional asset which would subsequently turn out to be stranded. Since conventional CBA does not model unexpected outcomes, the problem of stranding and the detail of its varying significance across competing technical solutions can fail to be fully captured by conventional CBA.

The academic subdiscipline of Real Options Analysis considers aspects of flexibility in real-world investment projects including the ability to delay, expand, or scale back the investment. These aspects of flexibility are all relevant to capital investments in electricity distribution networks. In addition, the advent of smart grid technologies adds competing choices between conventional and smart technical solutions to the same problem at the same network location. As discussed above, these competing technologies typically include both flexible and inflexible solutions, which should be compared on a common basis.

An additional consideration for electricity distribution networks is that the level of uncertainty in investment analyses is expected to increase over the coming years. This is due to the increasing presence in distribution networks of distributed energy resources such as rooftop solar photovoltaic panels and battery energy storage systems, as well as high-current consumer loads including those from electric vehicles. Uncertainty increases the risk of under-investing in distribution networks or, as described above, of investing in assets which may subsequently become stranded. The ultimate cost of such inefficient capital allocation is, of course, borne in large part by end-use consumers.

The research described in the present article demonstrates a bespoke probabilistic approach developed in the framework of Real Options Analysis which explicitly models randomness and flexibility. This approach thus enables the comparison of multiple competing investment strategies, including DSR and conventional investments. Further, it informs investment decision making through both economic cost and physical risk metrics. This approach to investment decision making is now in regular use at ENWL, providing a principled, quantitative basis for business cases for investment in smart grid technologies.

3 Informing Strategic Investment Decision-Making at ENWL

The collaboration between academia and industry described in the present article began with joint workshops which enabled the main ideas, together with important real-world practical and business considerations, to be shared. Based on this common understanding, the outline of a potentially appropriate ROA model was produced in report form. The parties subsequently agreed to proceed with a phase of work developing the approach into a prototype software tool.

In order to maximise the potential for rapid collaboration during the software development phase, the decision was taken to create the prototype model in the form of a spreadsheet. This prototype was completed following several further successful rounds of iteration and workshops, which refined both the approach and necessary implementation details. The resulting model was given the name RO CBA (Real options cost benefit analysis), and subsequently extended and refined by ENWL into a decision-making process and software tool with which to evaluate the value and impact of network investments.

In parallel to software development, the collaborators also worked to document their approach to ROA for electricity distribution networks in a peer reviewed journal article [4]. The latter work describes and justifies the novel stochastic model used in the RO CBA tool to describe the inherently random process of future peak network loads. Importantly the stochastic model does not assume the dynamics of geometric Brownian motion, an approach that is commonplace in mathematical finance, but one which has not been justified as a way of modelling electricity demand. Instead, RO CBA develops the network operator's own long-term demand forecasts into 'drift' functions for a stochastic process.

The article [4] provides a number of case study examples designed to maximise its potential impact. These include the problem of quantifying appropriate demand-side response compensation (customers participating in demand-side response require compensation, as such schemes rely upon customers temporarily reducing their demand). In particular the case studies illustrate the key practical finding that in the presence of uncertainty over future peak load growth, DSR can be an economical option to delay or even avoid large conventional network investments, thus reducing overall costs for both networks and their customers.

4 Influencing National Guidelines

While the present article describes work relating to electricity distribution networks, parallel considerations apply at the level of the national electricity transmission network. Due in large part to decarbonisation and the penetration of stochastic renewable generation, uncertainty is increasing at all levels of power systems globally. In the UK, National Grid plc owns and operates the electricity transmission system, and applies a methodology known as Network Options Assessment (NOA) to make annual investment recommendations based on a set of future demand scenarios.

In this context, in September 2016 the UK electricity and gas markets regulator Ofgem initiated a review of methodologies applied to investment decision-making by National Grid. The purpose of the review was to ensure that the most efficient capital investment recommendations would continue to be made by NOA in coming years. The present author provided verbal and written submissions to the review process, and the article [4] was cited in its final report [5]. These submissions demonstrated that implicit assumptions within the commonly used Black-Scholes framework were not appropriate to the network reinforcement context. Conversely the demonstrable value derived by ENWL from implementation of RO CBA, which applied Real

Options Analysis to model network investment risk, improving performance and saving customers money, led the NOA methodology review to recommend the approach described in [4] as an example of best practice. In this way, the work described in the present article has been instrumental in influencing the methodological review and the resulting recommendations for the inclusion of Real Options Analysis within National Grid's network investment analyses.

5 Enabling Better-Value Electricity Supply to Millions of Customers

ENWL is one of fourteen electricity distributors in the UK and serves 2.4 million of the UK's 30 million customers, in the north-west of England. It owns, operates, maintains and upgrades 13,000 km of overhead lines, 44,000 km of underground cables, and related transformers and switchgear. The RO CBA tool has been adopted by ENWL to compare investment strategies using a set of long-term demand forecasting scenarios. To date it has been used to make fourteen separate investment decisions, with over 100,000 customers directly benefitting from local upgrades across these projects.

The RO CBA tool has by now found multiple applications within ENWL beyond its intended purpose, for example in the planning of asset replacement schemes. Across its network, ENWL faces two key issues: insufficient network capacity (requiring network reinforcement), and poor network health (requiring asset replacement). While RO CBA was originally developed with the intention of comparing traditional network reinforcement strategies against innovative demand side response solutions to address network capacity issues, the model is also now being applied to asset replacement schemes in order to determine the most cost-effective replacement option. According to the company, this further application of the model has greatly expanded its reach and usefulness within ENWL, as there is approximately triple the number of replacement schemes versus reinforcement schemes.

ENWL recovers its costs by charging electricity suppliers for use of the network, which for domestic customers equates to about 16% of their average electricity bill. The RO CBA software tool provides a much improved method of risk assessment over the company's prior process, which previously was a time-consuming task performed by hand, which did not identify an optimal strategy and lacked additional decision metrics (such as risk assessments) that the RO CBA model provides. The RO CBA model enables the company to quantify the impact of each competing investment strategy, to ensure that the most cost-effective investments are made while appropriately managing risk.

Since these network costs are shared, financial savings are passed back to all 2.4 million of ENWL's customers in the form of reduced electricity bills. In this way the use of RO CBA has improved the cost effectiveness of electricity supply for 5 million people across 2.4 million homes and businesses.

References

1. Poudineh, R., Jamasb, T., 2014. Distributed generation, storage, demand response and energy efficiency as alternatives to grid capacity enhancement. Energy Policy **67**, 222–231.
2. Ofgem, 2013. Strategy decision for the RIIO-ED1 electricity distribution price control: Outputs, incentives and innovation.
3. Ofgem, 2012. Real options and investment decision making, consultation. London, UK.
4. Schachter, J. A., Mancarella, P., Moriarty, J. and Shaw, R., 2016. Flexible investment under uncertainty in smart distribution networks with demand side response: Assessment framework and practical implementation. Energy Policy **97**, 439–449.
5. National Grid Plc, 2017. Network Options Assessment Methodology Review.

Forecasting Gas Demand

Sarah E. Heaps, Kevin J. Wilson, and Malcolm Farrow

Abstract Efficient management of a gas distribution network requires the ability to forecast demand for gas over various timescales. Gas demand varies from day to day for a number of reasons, many of which can be included in a model. We describe a project to construct a new model for forecasting gas demand in two neighbouring regions of the UK. In particular we include a novel approach to modelling the effects of public holidays. We fit the model using Bayesian inference which allows the use of expert prior beliefs, flexibility in aspects of the model, and the production of forecasts with associated uncertainty measures which reflect uncertainty in model parameters as well as "random" variation. The results of the work are now used by a gas distribution company.

1 Introduction

The work described here is concerned with modelling and forecasting the demand for gas for use by customers. These customers include domestic customers who use the gas for heating and cooking as well as larger organisations who use the gas, for example, for heating large buildings and industrial users who use the gas in industrial processes, such as in furnaces or in electricity generation. In the UK, gas is moved long distances around the country in a national network which is owned and operated by National Grid. However the gas reaches the customers through regional

S. E. Heaps
Department of Mathematical Sciences, Durham University, Stockton Road, Durham DH1 3LE, UK
e-mail: sarah.e.heaps@durham.ac.uk

K. J. Wilson (✉) · M. Farrow
School of Mathematics, Statistics and Physics, Newcastle University, Newcastle upon Tyne NE1 7RU, UK
e-mail: kevin.wilson@newcastle.ac.uk

M. Farrow
e-mail: malcolm.farrow@newcastle.ac.uk

P. J. Aston (ed.), *More UK Success Stories in Industrial Mathematics*,
Mathematics in Industry 42, https://doi.org/10.1007/978-3-031-48683-8_30

gas distribution networks, owned and operated by other companies, which take the gas from the national network and deliver it to customers. This project involved a collaboration between one of these regional gas distribution companies, Northern Gas Networks (NGN), and Newcastle University. NGN has two distribution networks, known as local distribution zones (LDZs), the Northern LDZ and the North Eastern LDZ, both in the North of England.

The collaboration included three projects, involving respectively (1) modelling and forecasting daily demand for the whole of each LDZ, (2) hourly forecasting of demand on smaller parts of the network and (3) monitoring "alarms", that is signals received from remote sensors around the network indicating faults or problems. This article describes the first of these, which formed the basis of an Impact Case Study.

1.1 The Forecasting Problem

For demand forecasting purposes, customers are divided into four subsets, known as "load bands". Load bands 1–3 correspond to different ranges of gas demand, increasing from 1 to 3. Domestic customers fall in Load Band 1. Customers with commercial premises, for example, typically fall in Load Band 2. Load Band 3 represents, for example, small industrial customers. We will not consider Load Band 4 in this article. This represents large industrial customers with very high demand for gas. Demand in Load Band 4 behaves quite differently to that in Load Bands 1–3.

Demand forecasts, over a range of horizons, are required by both National Grid and the distribution companies for reasons including safety and security of supply and investment and operational planning [1]. The demand for gas is affected by many factors, which vary on different timescales. Long term trends in economic and social factors affect the annual total demand. Other factors and variables, such as the day of the week and the weather, affect the daily demand on much shorter timescales within this. The method that is used by distribution companies involves forecasts $\hat{A}_{u,h}$ for the annual demand $A_{u,h}$ in year u from a subset h of customers, i.e., a load band in an LDZ, $h = 1, \ldots, H$. These forecasts are produced by an external company. In the standard method, forecasts for demand $\tilde{Y}_{t,h}$ for subset h on day t are produced by multiplying these annual totals by scale factors specific to the day and subset. These products are then rescaled to match this annual forecast [1].

This project began when NGN expressed an interest in having a means of determining suitable scale factors, independent of the national system, which would be transparent to NGN staff and which could be altered to suit NGN requirements. In particular, there was an interest in representing the effects of public holidays on gas demand. Thus what was required was inference about a set of scale factors $\exp(k_{t,h})$ for the days of the year so that the forecast mean of $\tilde{Y}_{t,h}$, for example, would be $\mathrm{E}\{\exp(k_{t,h})\}\hat{A}_{u,h}$, since the scale factors may reasonably be treated as independent of the annual demand level. These scale factors should reflect a number of effects, including the effects of public holidays.

1.2 Outline of Our Proposed Solution

To address this problem, we proposed to develop a model for daily gas demand, using data from past years to learn about the values of model parameters and to examine the adequacy of the model. We proposed to use Bayesian inference for this task. It would then be possible to produce forecasts for daily gas demand and, as well as point forecasts, give realistic representations of the uncertainty around these point forecasts, in the form of predictive distributions. These predictive distributions would reflect both the *aleatory* uncertainty, that is the residual, unpredictable, "random" variation around the forecast mean, and the *epistemic* uncertainty, that is our remaining lack of knowledge about the precise values of model parameters.

In Sect. 2 we will outline our model and comment on the prior distribution. In Sect. 3 we will describe how we fitted the model and comment on the results. Further details of both of these, along with additional references, may be found in [2]. Finally, in Sect. 4 we will comment on the benefits of the work to NGN.

2 Model and Prior

As we used Bayesian inference, as well as a model, we required a prior distribution for the model parameters. The prior distribution reflects beliefs and uncertainties about model parameters before we take account of the data. It allows us to incorporate expert prior judgements and reasonable prior beliefs, for example, about similarities between quantities, without having to build fixed assumptions into the model.

2.1 Model

As well as calendar information, including the time of year, day of the week and whether a day was a public holiday, the variables in the model are, for each day and each LDZ, the total gas consumption for each of the three load bands and the value of the *Composite Weather Variable* (CWV) which is defined by National Grid. The CWV reflects air temperature but also takes into account other weather variables such as wind speed and involves some modifications at low and high temperatures to improve the linearity of the relationship with gas consumption [1].

Consider just one of the load bands. Let $\tilde{Y}_{t,j}/10$ be the gas demand, in gigawatt-hours, and let $w_{t,j}$ be the CWV value on day t in LDZ j. To help stabilize the variance and give additive effects in our model multiplicative effects on gas demand, we take logarithms. Let $Y_{t,j} = \ln(\tilde{Y}_{t,j})$, $\mathbf{Y}_t = (Y_{t,1}, Y_{t,2})'$ and $\mathbf{w}_t = (w_{t,1}, w_{t,2})'$.

Preliminary analysis showed that the residuals, after allowing for explanatory variables, were auto-correlated. A carry-over effect in gas demand data was also noted by [3] (Sect. 3.10.4). To reflect this, we use a first-order vector autoregressive, or VAR(1), process

$$\mathbf{Y}_t - \boldsymbol{\mu}_t = \Psi(\mathbf{Y}_{t-1} - \boldsymbol{\mu}_{t-1}) + \boldsymbol{\epsilon}_t, \qquad \boldsymbol{\epsilon}_t \sim \mathrm{N}_2(\mathbf{0}, \Omega_t^{-1}).$$

Here $\boldsymbol{\mu}_t$ is a mean vector which depends on the explanatory variables, Ψ is a 2×2 matrix with elements constrained to satisfy the stationarity conditions for a VAR(1) process (see [3]), $\mathbf{0}$ is a vector of zeroes and Ω_t is a 2×2 precision matrix, which depends on t. If we set $\Omega_t = \Omega$, then, given the appropriate initial distribution, $\mathbf{Y}_t - \boldsymbol{\mu}_t$ would follow a stationary VAR(1) process.

The elements of the mean vector $\boldsymbol{\mu}_t$ are given by

$$\mu_{t,j} = \alpha_j + B_{t,j,s_t}\beta_{j,r_t} + \Gamma_{t,j} + \Delta_{t,j} + (\zeta_{j,1} + \zeta_{j,2}w_{t,j})(w_{t,j} - m_{d(t),j}). \tag{1}$$

The intercept α_j varies between LDZs but not over time. The term $(\zeta_{j,1} + \zeta_{j,2}w_{t,j})(w_{t,j} - m_{d(t),j})$ represents the effect of the weather. The subscript $d(t) \in \{1, 2, \ldots, 366\}$ is the position in the year of day t and $m_{d(t),j}$ is a smoothed average for the CWV in LDZ j on day t. The two terms $\Gamma_{t,j}$ and $\Delta_{t,j}$ represent the effects of the day of the week and the annual seasonality respectively. Both of these are constructed using Fourier series. For $\Delta_{t,j}$ we use the maximum number, six, of unconstrained coefficients, while avoiding confounding, since the seven effects of days of the week sum to zero. So

$$\Delta_{t,j} = \sum_{k=1}^{3} \left\{ \delta_{j,1,k} \cos\left(\frac{2\pi kt}{7}\right) + \delta_{j,2,k} \sin\left(\frac{2\pi kt}{7}\right) \right\}.$$

The Fourier series for $\Gamma_{t,j}$ has period 365.25 and is truncated at K_γ harmonics, beyond which we assume that the contribution is negligible. For discussion of the choice of K_γ, see [2].

2.1.1 Public Holidays

The remaining term in (1), $B_{t,j,s_t}\beta_{j,r_t}$ represents the effects of public holidays. Public holidays, for example Christmas Day and Easter Monday, have different characteristics and occur at different times of year. To deal with this, we grouped the holidays into three "types." So $r_t \in \{1, 2, 3\}$ denotes the type of the closest holiday to day t (the following holiday in the event of a tie). In addition, the effect on gas demand is often not restricted to the day of the holiday itself. For example, if there is a holiday on a Monday, this is likely to affect the whole weekend and perhaps some following days as well. To deal with this we introduced the idea of *proximity days* for the days close to the holiday where demand is affected. Each day has a state $s_t \in \{1, 2, 3, 4\}$. These states are assigned to: (1) proximity days preceding a holiday, (2) holidays, (3) proximity days following a holiday, and (4) normal days, unaffected by holidays. Then

$$B_{t,j,1} = \rho_{\beta,j}^{n_t}, \quad B_{t,j,2} = 1, \quad B_{t,j,3} = \rho_{\beta,j}^{\min(n_t, p_t)}, \quad B_{t,j,4} = 0.$$

Here $\rho_{\beta,j} \in (0, 1)$ is a parameter governing the decay of the effect with increasing distance from the holiday, n_t is the number of days until the next holiday and p_t is the number of days since the last holiday. If day t is a holiday then $n_t = p_t = 0$.

We do not know how many proximity days there are before or after a holiday so we used a discrete stochastic model to describe the changes of state, with nonzero probabilities for the permitted transitions on a given day. The transition probabilities from state 4 to state 1 and from state 3 to state 4 depend on the values of n_t and p_t. Details are given in [2]. Of course we know when the process is in state 2, but the other states are unobserved.

Finally, the precision matrix Ω_t is allowed to depend on the state, with different values for states 2 (holiday) and 4 (normal). For states 1 and 3 the value is between these and depends on n_t and p_t. We also allow Ω_t to depend on the time of year, using a truncated Fourier series. To do this, we express Ω_t in terms of its square-root-free Cholesky decomposition to give a set of unconstrained parameters, which makes incorporation of the structure straightforward.

2.2 Prior

As well as a model we need a prior distribution for the unknown parameters. Space does not permit a full description here. Many parameters can be easily related to observable quantities which makes choosing prior distributions more straightforward. Clearly our prior beliefs will tend to give a positive correlation to a pair of corresponding parameters for the two LDZs. We therefore use hierarchical priors in such cases. One interesting aspect of the prior specification relates to beliefs about the numbers of proximity days before and after holidays. Prior distributions for the relevant parameters were tried and the prior predictive distributions examined until they appeared to represent the beliefs expressed by experienced NGN personnel.

3 Inference and Results

The model was initially fitted using data from a nine-year period preceding the project. Computation of summaries of the posterior distribution in such a model requires numerical methods. Fortunately efficient algorithms are available, usually based on taking many random samples from the posterior distribution. We used Hamiltonian Monte Carlo sampling [4] implemented in the Stan software [5].

We examined the fit of the model by comparing posterior predictive distributions with observed data and found that this was satisfactory. By comparing the results from our four-state model with those from a simpler, two-state, model with no proximity days, we found that the four-state model gave clear improvements for load bands 2 and 3. This improvement was not apparent for load band 1 where the inclusion of

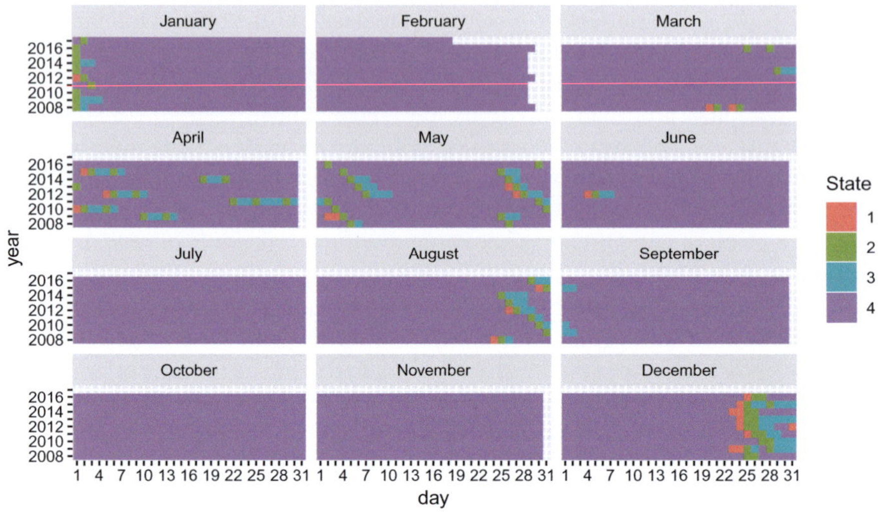

Fig. 1 The pointwise posterior mode for the state sequence, for load band 2

proximity days appeared to make little difference. Our approach led to forecast distributions, rather than simply point forecasts, and these forecast distributions compared well with observed demand data. Figure 1 provides the pointwise posterior mode for the state sequence s_t in load band 2 for the years 2008–2017. We see evidence to support the presence of all four states in the model and a pattern that is different for different public holidays, highlighting the flexibility of our approach.

Various modifications and improvements to the model are possible. However there are only nine Christmas Days in nine years of data so we cannot expect to make useful inferences about too many unknown parameters for the holiday effects. It was felt by NGN experts that patterns change over time and using too many past years would be inadvisable.

4 Benefits to NGN

Scale factors resulting from our model fit have been used by NGN in their annual medium-term forecasting exercise. We have repeated the model fitting every two years with more recent data to provide updated scale factors.

The work has been considered to be a success by NGN with improved ability to match supply to demand leading to considerable financial savings, including during periods of extreme weather. It has been possible to improve the scheduling of pipeline inspections. The model and forecasting has improved understanding of the behaviour of the system and helped NGN staff to make informed decisions.

Acknowledgements The authors thank NGN for providing the data and for many helpful discussions and comments. We are also grateful to Matt Linsley, NU Solve Manager at Newcastle University.

References

1. National Grid, *Gas Demand Forecasting Methodology*. (National Grid, Warwick, 2016)
2. Heaps, S.E., Farrow, M., Wilson, K.J.: Identifying the effect of public holidays on daily demand for gas. J. R. Statist. Soc. A **183**, 471–492 (2020)
3. Tunnicliffe-Wilson, G., Reale, M., Haywood, J., *Models for Dependent Time Series*. (CRC Press, Boca Raton, 2016)
4. Neal, R. M.: MCMC using Hamiltonian dynamics. In: Brooks, S., Gelman, A., Jones, G., Meng, X.-L. (eds.) Handbook of Markov Chain Monte Carlo, pp. 113–162. Chapman and Hall–CRC, Boca Raton (2011)
5. Carpenter, B., Gelman, A., Hoffman, M. D., Lee, D., Goodrich, B., Betancourt, M., Brubaker, M. A., Guo, J., Li, P. and Riddell, A.: Stan: a probabilistic programming language. J. Statist. Softwr., **76**, 1–32 (2017)

Food

Global Food Solutions Via Optimal Diet Formulations

J. A. J. Hall, K. I. M. McKinnon, A. Grothey, and I. L. Galabova

Abstract Format Solutions is the world's leading supplier of software for food formulation, and this requires high performance optimization software. Hall and colleagues have been supplying solver technology to Format for 25 years, enabling them to achieve a technological lead in the global food formulation software industry. This article sets out the story of the collaboration and its impact.

1 Introduction

Format Solutions is the world's leading supplier of food formulation software, and its competitive position depends on techniques and software developed by the University of Edinburgh Optimization group. This software is used to identify raw material blends and production strategies for companies responsible for more than half of the \$50Bn petfood market. With the solution of optimization problems being the overwhelmingly dominant computational cost of Format Solutions' software, post-2000 research and development by the Edinburgh group has led to its performance improving due to software alone by more than an order of magnitude.

1.1 Food Formulation Problems

The "Diet Problem" is a classical example used to illustrate the class of problems that can be formulated and solved via linear programming (LP). Given the unit costs and nutritional properties of a range of foodstuffs, the diet problem is to identify the cheapest combination of foodstuffs that satisfies daily nutritional requirements. Indeed, in the autumn of 1947, Stigler's original formulation was chosen to be the first practical

J. A. J. Hall (✉) · K. I. M. McKinnon · A. Grothey · I. L. Galabova
School of Mathematics and Maxwell Institute for Mathematical Sciences, University of Edinburgh, James Clerk Maxwell Building, Peter Guthrie Tait Road, Edinburgh EH9 3FD, UK
e-mail: jajhall@ed.ac.uk

P. J. Aston (ed.), *More UK Success Stories in Industrial Mathematics*,
Mathematics in Industry 42, https://doi.org/10.1007/978-3-031-48683-8_31

test of Dantzig's newly-invented "simplex algorithm" for solving LP problems. With 77 foodstuffs and nine nutritional constraints, 62 foodstuffs were eliminated due to being dominated in cost and nutritionally by the other 15, and the resulting problem was solved using mechanical calculators by nine people in two weeks.

Although the individual human diet problem is of relatively little commercial interest, the blending of simple raw materials to manufacture food is an industry that is worth approximately \$100Bn per annum. This is dominated by the petfood market, where small percentage savings are significant. To respond to variability in price and supply of raw materials, food producers must determine the optimal composition for daily manufacture and future planning.

Food formulation problem sizes increase with the number of recipes and raw materials, and larger companies wish to consider raw material usage over a number of mills. A secondary, but potentially very profitable enterprise, is to consider substituting materials when it is more valuable to trade ownership or purchase options for them on global commodity markets. To obtain the information for all this manufacturing and economic activity requires the solution of large-scale LP problems.

1.2 Mathematical Formulation and Solution of LP Problems

Mathematically, decisions to be made optimally are modelled by algebraic unknowns referred to as variables, and the values that the variables can take are normally required to lie between bounds. In the diet problem and many other applications, the cost of the decision is a linear combination of the variables, and general constraints on the decisions require linear combinations of the variables to lie between bounds. Formally, the LP problem may be written as

$$\text{minimize } f = \boldsymbol{c}^T\boldsymbol{x} \quad \text{subject to } A\boldsymbol{x} = \boldsymbol{b}, \quad \boldsymbol{x} \geq \boldsymbol{0}. \tag{1}$$

LP problems are the fundamental model for optimal decision-making and Dantzig's simplex algorithm remains the more important of the two approaches used to solve them. Indeed, such has been its impact that it was judged to be one of the top ten algorithms of the twentieth century [1]. Thus, for decades, the need to solve LPs efficiently drove research into enhancements of the simplex algorithm and computational techniques for its implementation. For practical LP problems, it is typical for most of the entries of the coefficient matrix A in (1) to be zero. As such, the matrix and LP are referred to as being "sparse", and exploiting the existence of these zeros when solving the LP is essential for computational efficiency.

The constraint matrix for a small feed formulation problem is illustrated in Fig. 1. Each diagonal block corresponds to the variables and constraints for a particular diet, and the first few rows of the matrix correspond to "linking constraints" to ensure that the total use of raw materials across all diets does not exceed the available supply. Without these linking constraints the LP would be a set of independent

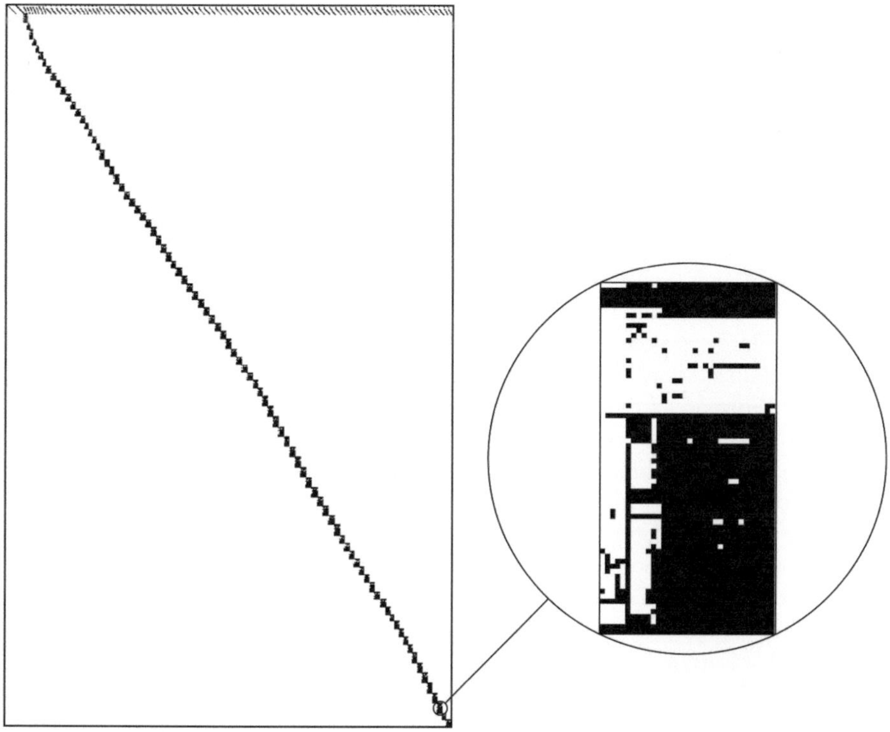

Fig. 1 Nonzero locations and example block structure of a multi-diet feed formulation problem with 3683 variables, 2068 constraints and 44651 nonzeros

individual diets. Although more than half of the entries in a particular diagonal block are nonzero, only 0.6% of the matrix entries are nonzero. If this sparsity is not exploited, then most matrix computations would involve addition or multiplication by zero. Although many researchers had tried to develop simplex implementations to exploit parallel computers when solving general large-scale sparse LP problems, very little of practical value had been achieved by 2010 [2]. These attempts included the work of Hall and McKinnon, a by-product of which was `EMSOL`, an efficient simplex solver written in `FORTRAN`.

2 Collaboration with Format

Format International was a Woking-based company that combined the animal nutrition expertise of Merryl Webster with the computing and IT skills of John Hall to produce software for optimal animal feed formulation. In 1997, being aware of the inefficiency of John's simplex solver, and having learned through a mutual friend that LP solver experience was available in Edinburgh, they visited to ask for advice. EMSOL's superiority was soon clear, so Format bought a copy, and the relationship between Format and the Edinburgh group was established. Since the solution of LP problems dominates the computational requirements of Format's software, adopting EMSOL had a significant impact on Format's growth, enabling it to win new business with larger clients.

Within a couple of years, Format benefitted from a major performance enhancement in EMSOL as a consequence of Hall and McKinnon's prize-winning paper [3] on identifying and exploiting the phenomenon of hyper-sparsity when solving LP problems using the simplex algorithm. As a result, the solution time using EMSOL for Format's larger problems improved by a factor of more than five, and the cost of post-optimal analysis was reduced by a factor of more than eight.

In 1999, Format set out a problem that one of their major customers wanted to solve, in which production requirements dictate that raw materials are used to produce several pre-mixed combinations, before final blending into products. Such problems are often found in industries such as pet food and dried milk production. Mathematically, this "pooling" problem is fundamentally more complex, in that the formulation involves products of decision variables. As such it is nonlinear and non-convex, so could not be solved using the software available to Format. So, in 2000, Grothey and McKinnon [4] developed and implemented a bespoke technique for the pooling problem. This required the solution of a sequence of LP problems (using EMSOL) and was supplied to Format in early 2001. The ability to solve pooling problems gave Format a unique position within the feed formulation industry, allowing them to secure a contract with a manufacturer producing 60% of the world's pet food.

In 2015, Cargill Inc., one of the world's leading animal feed manufacturers, and a long-time customer of Format, acquired Format International, renaming it Format Solutions. This purchase was made following a review of the future direction of their formulation software solutions across the world-wide business. The Cargill review identified that Format's own software developments, coupled with its optimization capabilities and the talent in the Format team were far ahead of Cargill's own in this area, and to match them would have taken many years and much expense. The purchase of Format International, bringing with it the ability to acquire the software, its IP and solvers, was assessed as a far better option than developing their own solution.

3 Enhanced Solvers for Format and the World

In 2009, Hall recommenced research into high performance implementations of the simplex algorithm. His PhD student, Qi Huangfu, was partially funded by Format to work in the area, leading to the development of serial and parallel implementations of the simplex algorithm in C++ that offered significant performance improvements over EMSOL and, academically, represent the state of the art [5]. When Huangfu moved to FICO inc. in 2013, incorporation of his work yielded the world's only parallel commercial simplex solver, and contributed meaningfully to it topping the optimization industry's independent benchmarks for most of the next five years.

The value to Format of Huangfu's work was realised in 2016 when Galabova joined Hall to study for a PhD, and implemented pre-processing techniques to reduce the dimension of the LP to be solved. These generalisations and extensions of the very first presolve techniques applied to Stigler's diet problem had not been implemented in `EMSOL`. She then combined them with Huangfu's simplex solver and demonstrated a performance improvement of almost ten over `EMSOL` for Format's larger problems. Impressed, Format funded further solver developments that led to the creation by Hall and Galabova of `HiGHS`. This open-source linear optimization software system has been widely embraced by both users and interface developers and, based on the independent benchmarks, is now the world's best. As well as enhancing the solvers available to Format, `HiGHS` is already creating impact in many other application areas.

A new chapter in the story began in 2022, when Datacor bought Format Solutions from Cargill. The Edinburgh team looks forward to continued interaction in this new environment, with hopes that `HiGHS` can add to its portfolio of impact!

References

1. B. A. Cipra. The best of the 20th century: Editors name top 10 algorithms. *SIAM news*, 33(4):1–2, 2000.
2. J. A. J. Hall. Towards a practical parallelisation of the simplex method. *Computational Management Science*, 7(2):139–170, 2010.
3. J. A. J. Hall and K. I. M. McKinnon. Hyper-sparsity in the revised simplex method and how to exploit it. *Computational Optimization and Applications*, 32(3):259–283, 2005.
4. A. Grothey and K. I. M. McKinnon. On the effectiveness of sequential linear programming for the pooling problem. *arXiv preprint* arXiv:2002.10899, 2020.
5. Q. Huangfu and J. A. J. Hall. Parallelizing the dual revised simplex method. *Mathematical Programming Computation*, 10(1):119–142, 2018.

Combating Crop Losses and Improving Global Food Supplies Through Mathematical Modelling of "Gene Silencing"

Konstantin B. Blyuss and Yuliya N. Kyrychko

Abstract One of the biggest challenges facing humanity is the need to maintain adequate food supplies in the face of growing world population. To achieve this, we need to have effective tools protecting crops against a variety of parasites that can cause crop losses. While pesticides are the traditional means of addressing this problem, with time they become less effective due to parasite adaptation, and also we become more conscious of their potential negative impact on the environment, as well as on human health. One very promising technology that can provide an effective and safe approach to protecting plants against parasites is RNA interference, also known as "gene silencing". This work has used mathematical models to establish strategies for optimal use of nutrients and pesticides that improve plant growth and protect plants against parasites, while also minimising pesticide use. Mathematical models of RNA interference have helped understand main features of this process, and this was then used for the development of natural microbial biostimulants that, besides improving crop performance, provide a targeted protection against specific parasites, while avoiding detrimental effects on the environment.

1 Introduction

The last few decades have witnessed an enormous growth of the world population that has increased from 6 billion people in 1999 to almost 8 billion people in 2020. This has brought into focus the need to sustainably grow and produce huge amounts of various crops to satisfy global demand. While a variety of different countries and regions have been involved in crop farming over the years, they continue to face challenges of different origin, including climate change, as well as plant pests and parasites. They cause huge economic losses, while also putting at risk livelihoods

K. B. Blyuss (✉) · Y. N. Kyrychko
Department of Mathematics, University of Sussex, Falmer BN1 9QH, UK
e-mail: k.blyuss@sussex.ac.uk

Y. N. Kyrychko
e-mail: y.kyrychko@sussex.ac.uk

P. J. Aston (ed.), *More UK Success Stories in Industrial Mathematics*,
Mathematics in Industry 42, https://doi.org/10.1007/978-3-031-48683-8_32

of communities that rely on farming as their main source of income, not to mention threatening their food resources. Whereas a number of different tools have been used over the years to minimise crop losses, in order to reduce negative effects of pesticides and limit associated costs, recently attention has been drawn to more environmentally friendly and organic approaches to crop protection. In this project we have looked at how mathematical models can be used to improve crop protection against parasites, thus helping reduce losses and maintain sustainable food production.

There are two distinct components in our work. The first part focused on investigating how farmer awareness about the degree to which crop plants are affected by certain plant viruses, can be used to apply nutrients and insecticides that facilitate plant growth and protect plants against further infections. The second aspect of our work concerns using RNA interference (RNAi), also known as "gene silencing", to develop natural microbial biostimulants that protect plants against parasites. The main mechanism of RNA interference lies in cells' ability to stop protein production from genes that have already been transcribed. This is a fundamental biological property of all eukaryotic cells, i.e. cells having a nucleus, and its discovery was recognised by a Nobel Prize in 2006. For the particular purpose of protecting plants against parasites and infections, RNAi can be used in two different ways. On the one hand, it can protect plant cells from becoming infected by plant viruses through silencing genes necessary for viral replication, while also clearing viral genome from already infected cells, and protecting other cells from becoming infected. On the other hand, it can be used through a "Trojan horse"-like strategy to target parasites themselves: through root soaking or spraying, plants are provided with necessary RNA constructs, and when these constructs are ingested by the parasites feeding on those plants, this results in reducing parasites' fecundity or causing their death by silencing some of their essential genes.

As a testbed for applying our theoretical results, we have chosen Ukraine, due to its known standing as the world's leading global producer of sunflower, as well as being in top five producers of a large number of other staple crops, including wheat, barley, corn and potatoes. A recent military invasion of Ukraine by Russia has highlighted Ukraine's prominence on the global stage as a major producer of various food crops, and the war started by Russia has led to significant threats to communities and whole countries in various parts of the world, from Africa to Asia. The second reason for choosing Ukraine was the fact that scientists from the National Academy of Sciences of Ukraine had already been working for a number of years on understanding complex interactions between plants and their parasites, as well as the ability of various soil bacteria to produce natural compounds with anti-parasitic properties.

2 Mathematical Modelling of Using Nutrients and Insecticides

In order to understand how one can optimise the use of plant nutrients and insecticides for plant growth and protection depending on farmers' awareness of infection, we have developed models looking at controlling viral infections of plants caused by viral vectors [1, 2]. We considered a population of plants that were divided into healthy, x, latently infected l, and infected y, whereas the population of infected vectors is denoted by v. In the absence of infection, plants reproduce logistically, with the linear growth rate r and carrying capacity K. At rate λ uninfected plants are infected by vectors, thus becoming latently infected, and after an average latency period of $1/\alpha$ they become infected. Population of infected vectors grows at rate b proportionally to the number of infected plants, and they have an average lifetime of $1/\eta$. The level of farmers' awareness about the infection is denoted by M, and it grows with global awareness campaigns at rate ω_0, with observation of infected plants at rate σ, and dissipates at rate τ. Depending on the level of awareness, farmers apply nutrients at level $N(t) = \mu + \rho M/(1 + M)$, where μ is the baseline rate in the absence of any information, and ρ is the maximum rate, at which nutrients are applied. The effect of nutrients is twofold: they improve plant growth by increasing their linear growth rate $r(N) = r_0[1 + k_1 N(1 - k_2 N)]$ (nutrients are beneficial in small quantities but become harmful/toxic when applied in large amounts), and also nutrients reduce the capacity of plant virus to infect healthy plants, which is represented by a reduced infection rate $\lambda(N) = \lambda_0 \left[1 - k_3 \exp(-k_4 N)\right]$, $0 \le k_3 < 1$. Finally, besides nutrients, depending on their awareness M, farmers can also use insecticides at a rate $P(M) = \epsilon M/(1 + M)$ that cause the death of viral vectors. With these notations, the model for plant-vector interactions with awareness-based use of nutrients and insecticides has the form [1]

$$\begin{aligned} &\frac{dx}{dt} = r(M)x\left(1 - \frac{x + l + y}{K}\right) - \lambda(M)xv, \qquad \frac{dl}{dt} = \lambda(M)xv - \alpha l, \\ &\frac{dy}{dt} = \alpha l - my, \qquad \frac{dv}{dt} = by - \eta v - P(M)v, \qquad \frac{dM}{dt} = \omega_0 + \sigma y - \tau M. \end{aligned} \tag{1}$$

From the perspective of disease control, the most important role is played by the disease-free steady state of this model, which is given by $E_0 = (K, 0, 0, 0, \omega_0/\tau)$. If we define the basic reproduction number $\mathcal{R}_0$ as

$$\mathcal{R}_0 = \frac{bK\lambda(\omega_0/\tau)}{m\left[\eta + P(\omega_0/\tau)\right]} = \frac{bK(\tau + \omega_0)\lambda_0}{m[\eta\tau + \omega_0(\epsilon + \eta)]} \cdot \frac{1}{1 - k_3 \exp\left[-\frac{k_4[\mu\tau + (\mu+\rho)\omega_0]}{\omega_0 + \tau}\right]},$$

then we can show that the disease-free steady state will be stable, i.e. the disease will be eradicated, if $\mathcal{R}_0 < 1$. This provides a practically useful condition on various parameters, such as rates of application of nutrients and insecticides that are needed

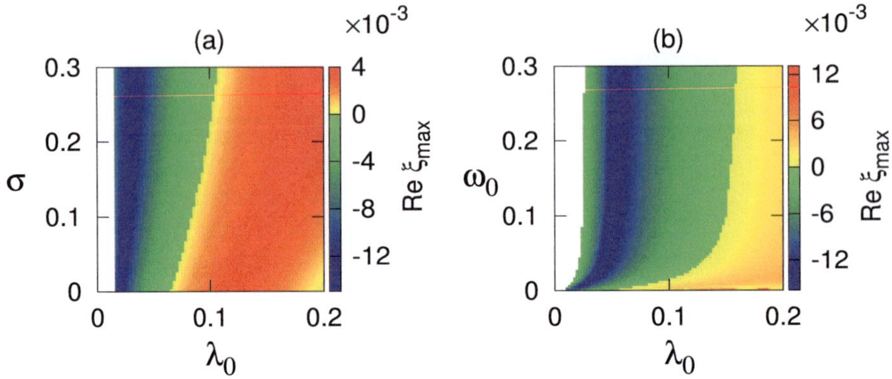

Fig. 1 Stability of the endemic steady state E^* of the model (1). Colour code denotes maximum real part of the leading characteristic eigenvalue whenever the endemic steady state is feasible. Parameter values are as follows, $r_0 = 0.05$, $k_1 = k_2 = k_3 = k_4 = 0.5$, $\mu = 0.5$, $\rho = 0.5$, $K = 1$, $\alpha = 0.1$, $b = 0.8$, $\zeta = 0.05$, $\epsilon = 0.25$, $m = 0.1$, $s = 0.6$, $\sigma = 0.05$, $\eta = 0.05$, $\tau = 0.016$. **a** $\omega_0 = 0.003$, **b** $\sigma = 0.015$.). Figure taken from Basir et al. [1], reprinted with permission

to eliminate plant infection. One important and unexpected conclusion from this result is that once the disease is endemic, it is impossible to eradicate it if awareness about plant infection increases only on account of observing infected plants, as shown in Fig. 1. This can be done, however, once awareness also increases due to global media campaigns, which would result the higher level of using insecticides. In the subsequent work, we have looked into how one can achieve a similar result by using natural microbial stimulants that simultaneously improve plant growth, protect them against disease, and cause mortality of vectors [2].

3 Mathematical Modelling of RNA Interference (RNAi)

To model interactions between plant cells and viral infection, as mediated by RNAi, we considered a single plant, and divided all plant cells into susceptible cells S, infected cells I that spread further infection within plant, and warned cells W that are able to withstand viral infection. Warned cells are the susceptible cells that have received silencing signal from infected cells and changed their cellular machinery in such a way that they can no longer be infected by the virus. Transmission rate of viral infection between infected and susceptible cells is given by parameter λ, recovery rate of infected cells is σ, while mortality rates of uninfected and infected cells are ϵ and z, respectively, with $z > \epsilon$. The effect of RNAi is twofold: it helps plant cells already infected by a virus recover faster, and it also provides a warning signal to other susceptible cells, thus reducing the overall infection. RNA signal from infected cells is assumed to spread at rate δ, causing susceptible cells to become warned, and causing infected cells to recover with rate ϕ through amplification of this silencing signal. New proliferating cells are produced at rate k by susceptible and warned plant cells. There are also two time delays in the model: τ_1 represents an average time it takes for proliferating cells to mature and become susceptible, while τ_2 is the time during which cells remain infected before the propagating RNA signal reaches other cells. With these assumptions, the model of plant immune response has the form [3]

$$\frac{dS}{dt} = k[S(t-\tau_1) + W(t-\tau_1)]e^{-\epsilon\tau_1} - S(t)[\lambda I(t) + \delta I(t-\tau_2) + \epsilon S(t)],$$

$$\frac{dI}{dt} = I(t)[\lambda S(t) - (z+\sigma) - \delta\phi I(t-\tau_2)],$$

$$\frac{dW}{dt} = \delta S(t)I(t-\tau_2) - \epsilon W(t).$$

The disease-free steady state of this model as given by $E_0 = \left(ke^{-\epsilon\tau_1}/\epsilon, 0, 0\right)$ is stable for all maturation periods, provided the rate of production of new proliferating cells k is smaller than some $k_{min} = \epsilon(\sigma + z)/\lambda$, while for $k > k_{min}$, this steady state is stable for sufficiently large maturation times $\tau_1 > \tau_{min}$ and unstable for $\tau_1 < \tau_{min}$, where $\tau_{min} = [\ln(k) - \ln(k_{min})]/\epsilon$. This result illustrates the role of proliferation rate and maturation time in controlling the infection. Numerical bifurcation analysis of this model has shown the role of other parameters in determining existence and stability of the disease-free and endemic steady state, which corresponds to a persistent plant infection. The results indicate that a higher rate of propagation and/or higher amplification factor of RNA signal result in the destabilisation of endemic steady state, as illustrated in Fig. 2. At the same time, a faster spreading RNA signal can only help infected plant to recover faster, but it cannot by itself lead to recovery. A faster propagating RNA signals can also provide a resistance phenotype, where initial infection is very quickly contained do the prevention of its subsequent spread, as opposed to a recovery phenotype, where significant parts of the plants first become infected.

In a subsequent paper [4] we focused on within-cell dynamics of RNAi process and showed how primed amplification can enhance RNA silencing. We also explored the role of time delays associated with the processes of production of new double stranded RNA, which is the main component of RNAi, from mRNA and garbage RNA. Analytical and numerical bifurcation analyses have shown that the system is able to support bi-stability between steady states with high and low mRNA concentrations, as well as between periodic orbits around these states when they are unstable.

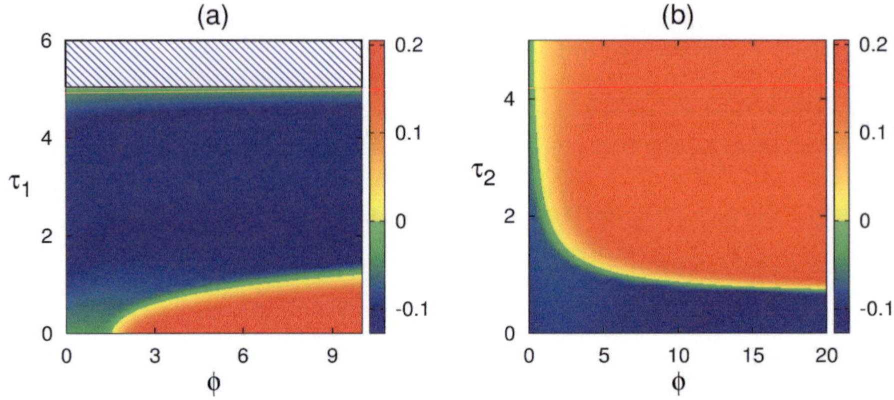

Fig. 2 Stability of the endemic and disease-free steady states with parameter values from Table 1. Diagonal blue indicates the region where the disease-free steady state is asymptotically stable, and the endemic steady state is not feasible. Colour code denotes real part of characteristic eigenvalues for the endemic steady state when it is feasible. Figure taken from Neofytou et al. (2018) [3], reprinted with permission

4 Impact of the Mathematical Models

Jointly with colleagues from Dnipro State Agrarian and Economic University in Ukraine, we translated our theoretical findings into guidelines for farmers that contain practical recommendations on using pesticides, nutrients and microbial biostimulants to improve plant growth and protect crops against parasites and pests. These recommendations set out concrete levels of applying nutrients/pesticides for different agricultural crops depending on the level of crop infection measured through observation. The institution we collaborated with is located in Prydniprovsky region of Ukraine, which is the largest steppe region that occupies a third of the country's steppe, and contains almost 20% of the total Ukrainian sown area of agricultural crops. The region is known for an abundance of fertile chernozem soil, and it alone produces over 20% of the total Ukrainian harvest of wheat, almost 22% of barley, and over 26% of sunflower. In fact, in 2019 that region alone produced 8% of the total amount of sunflower produced worldwide. Once the recommendations were introduced and distributed to local farmers, 40% of whom adopted them in their farming practice, this has led to an increase of 8–12% in crop viability, accompanied by a reduction of between 15% and 20% in costs associated with using nutrients and pesticides, because now they were applied more systematically depending on the observed level of crop infestation.

We have also established contacts with the Institute of Microbiology and Virology of the National Academy of Sciences of Ukraine, who for a number of decades had been working on understanding soil-plant interactions, with a particular emphasis on soil bacteria and their metabolites that can be beneficial for plants. Using the results of our models of RNA interference, and in particular, how this process develops in plants, and how it can be used to target plant parasites, colleagues in Ukraine selected and tested metabolites of soil bacteria that result in the highest levels of within-plant production of small interfering RNAs protecting plants against parasites. These metabolites were then used to produce natural microbial biostimulants that were tested under a variety of laboratory, greenhouse and field conditions for different crop plants, and we reported our findings in a joint publication [5]. Since the biostimulants were produced from natural products of soil bacteria, they can also be used for organic farming, which is a very significant achievement because organic farmers cannot use conventional pesticides to protect their crops against pests, and, as a result, they often suffer substantial crop losses. Moreover, since these microbial biostimulants are targeting specific genes of individual parasites, it means they are safe to other wildlife, in particular bees, whose populations have declined over the last several years due to the use of neonicotinoid pesticides. For this reason, our work received a very significant media attention from general and specialised farming publications in different languages on all continents.

Several Ukrainian companies have subsequently established industrial production of those natural microbial biostimulants, selling them to agricultural businesses in Ukraine and Kazakhstan (naturally, certification and licensing take considerable amount of time). The results of their application in large-scale farming practice are really impressive, with increased yields in sunflower, wheat and barley of 19–63%, 11–21% and 9–18%, respectively. When the same biostimulants were used by farmers growing vegetables, they have observed the increases of 10–12%, 20–66% and 23–26% in their yields of potatoes, tomatoes and cucumbers. One of these biostimulants has received an international "Organic standard" certification, and it is currently used by some of the largest organic farming companies in Ukraine. As mentioned earlier, with Ukraine being a major producer and exporter of these foods at international stage, these increased yields have benefited not only the companies involved in the production of corresponding crops, but also the global food markets and communities that rely on those crops as their staple foods.

References

1. Al Basir, F., Blyuss, K.B., Ray, S.: Modelling the effects of awareness-based interventions to control the mosaic disease of Jatropha curcas. Ecol. Compl. **36**, 92–100 (2018)
2. Blyuss, K.B., Al Basir, F., Tsygankova, V.A., Biliavska, L.O., Iutynska, G.O. et al.: Control of mosaic disease using microbial biostimulants: insights from mathematical modelling. Ric. di Matem. **69**, 437–455 (2020)
3. Neofytou, G., Kyrychko, Y.N., Blyuss, K.B.: Time-delayed model of immune response in plants. J. Theor. Biol. **389**, 28–39 (2016)
4. Neofytou, G., Kyrychko, Y.N., Blyuss, K.B.: Time-delayed model of RNA interference. Ecol. Compl. **30**, 11–25 (2017)
5. Blyuss, K.B., Fatehi, F., Tsygankova, V.A., Biliavska, L.O., Iutynska, G.O., Yemets, A.I., Blume, Ya.B.: RNAi-based biocontrol of wheat nematodes using natural poly-component biostimulants. Front. Plant Sci. **10**, 483 (2019)

Miscellaneous

Points of Convergence—Music Meets Mathematics

Lasse Rempe

Abstract *Phase-locking* is a fundamental phenomenon in which coupled or periodically forced oscillators synchronise. The *Arnold family* of circle maps, which describes a forced oscillator, is the simplest mathematical model of phase-locking and has been studied intensively since its introduction in the 1960s. The family exhibits regions of parameter space where phase-locking phenomena can be observed. A long-standing question asked whether *hyperbolic* parameters–those whose behaviour is dominated by periodic attractors, and which are therefore stable under perturbation–are dense within the family. A positive answer was given in 2015 by van Strien and the author, which implies that, no matter how chaotic a map within the family may behave, there are always systems with stable behaviour nearby. This research was a focal point of a pioneering collaboration with composer Emily Howard, commencing with Howard's residency in Liverpool's mathematics department in 2015. The collaboration generated impacts on creativity, culture and society, including several musical works by Howard, and lasting influence on artistic practice through a first-of-its-kind centre for science and music. We describe the research and the collaboration, and reflect on the factors that contributed to the latter's success.

1 Introduction

In the 17th century, the Dutch scientist Christiaan Huygens discovered that two pendulum clocks, coupled by being mounted on the same wooden beam, synchronised their movements. This phenomenon, described by Huygens as "a miraculous sympathy," is called *phase locking* (also mode locking, or entrainment): interacting periodic oscillators tend to synchronise their movements (with the same period or with periods that are related by an integer multiple). Phase locking is near-ubiquitous in physical oscillators: Examples include the fact that the revolution period and orbital period of the moon are identical and the synchronisation of fireflies. A similar

L. Rempe (✉)
Department of Mathematics, The University of Manchester, Oxford Road, Manchester M13 9PL, UK
e-mail: Lasse.rempe@manchester.ac.uk

P. J. Aston (ed.), *More UK Success Stories in Industrial Mathematics*,
Mathematics in Industry 42, https://doi.org/10.1007/978-3-031-48683-8_33

effect occurs when one oscillator is *forced* by another, for example in bowed musical instruments: When a violin string is plucked by hand, it exhibits non-harmonic overtones; when bowed, all of these overtones are forced onto the harmonic series.

A goal of "pure" (i.e., theoretical) mathematics is to investigate interesting phenomena in their most fundamental settings, to understand the underlying principles. In a 1961 paper [1, Sect. 12], Vladimir Arnold introduced what may be the simplest model for phase-locking behaviour: a family of self-maps of the circle, motivated by the movement (in discrete time-steps) of a forced periodic oscillator. Known as the *Arnold family* of circle maps, it can be described by the formula

$$f_{\alpha,b}(\theta) := \theta + \alpha + b\sin(\theta) \pmod{2\pi}. \tag{1}$$

The angle θ represents a point on the circle $S^1 = \mathbb{R}/2\pi\mathbb{Z}$. The number α is a rotation parameter (also an element of S^1); if $b = 0$, then $f_{a,b}$ is simply the rotation by an angle of α. The final term in (1) is a forcing term, which must also be a function of θ, and thus periodic. We use the simplest possible periodic forcing term, namely a multiple of sin.[1] The parameter $b \in [0, \infty)$ determines the strength of the forcing.

Let us fix parameters α and b and write $f = f_{\alpha,b}$. Given a starting state $\theta_0 \in \mathbb{R}/2\pi\mathbb{Z}$, we think of $\theta_1 = f(\theta_0)$ as determining the state of our dynamical system after one time step. So $\theta_2 = f(\theta_1)$ is the state after two steps, and after n steps:

$$\theta_n = f(\theta_{n-1}) = f(f(\ldots f(\theta_0)\ldots)) \quad (n \text{ times}).$$

The sequence $(\theta_n)_{n=0}^{\infty}$ is called the *orbit* of θ_0 under f. If $\theta_n = \theta_0$ for some $n > 0$, then the (finite) orbit is called a *periodic cycle*; this cycle is *stable* (under perturbation of θ_0) if the orbits of all nearby starting values converge to it.

For $b < 1$ (the region below the grey line in Fig. 1), the map $f_{\alpha,b}$ is a diffeomorphism of the circle. It indeed exhibits phase-locking phenomena, which are well-understood: For parameters in certain regions, called "Arnold tongues" (the white regions), all but finitely many orbits tend to a stable periodic cycle. Moreover, these maps are stable under perturbations of f within the Arnold family: a small change of the parameter leads to a small change in the overall behaviour of orbits, so $f_{\alpha,b}$ is "phase-locked" to the periodic orbit.

For $b > 1$, the map $f_{\alpha,b}$ is no longer invertible and has two distinct critical points on S^1. The Arnold tongues begin to intersect, and the periodic orbit for a given tongue may bifurcate and become unstable. This can lead to *chaotic* phenomena, where an arbitrarily small change of the starting state θ_0 could lead to completely different long-term outcomes. It is thus natural to ask whether for $b > 1$, there is still a dense set of parameters whose behaviour is stable under perturbations of the parameter.

More precisely, $f = f_{\alpha,b}$ with $b > 1$ is called *hyperbolic* if the orbits of both critical points tend to stable periodic cycles. Such f is stable under perturbations of the parameter–all nearby maps are also hyperbolic–and almost every orbit converges to a stable cycle. See Fig. 1, where white regions correspond to hyperbolic

[1] Arnold used cos instead of sin, which is equivalent.

Fig. 1 Parameter space of the Arnold family (1) (where $\alpha \in (-\pi, \pi)$ and $b \in (0, 4)$)

maps. Density of hyperbolic maps in parameter spaces is a central question of one-dimensional dynamics. For real polyomials, it was established by Kozlovski, Shen and van Strien [2] in 2007, answering part (b) of Smale's 11th problem. However, this does not resolve the question in the Arnold family: a key issue is that f, when extended to the complex plane, is transcendental rather than algebraic (see below). This problem was overcome by van Strien and the author [3][2]:

Theorem 1 (Density of hyperbolicity) *Hyperbolic maps are dense in the Arnold family. That is, given any $\alpha \in S^1$ and $b > 1$, and every $\varepsilon > 0$, there exist perturbed parameters $\alpha' \in S^1$ and $b' > 1$ with $|\alpha - \alpha'|, |b - b'| < \varepsilon$ such that $f_{\alpha',b'}$ is hyperbolic.*

Theorem 1 states that the white regions in Fig. 1 are dense; the apparent presence of solid black regions arises from the fact that stable regions may be very thin.

2 Ideas in the Proof of the Theorem

To prove Theorem 1, we consider the *complex extension* of $f_{\alpha,b}$, allowing the state θ in (1) to be a complex number. Applying the change of variable $z = \exp(i\theta)$, we obtain a self-map of the punctured plane $\mathbb{C}^* = \mathbb{C} \setminus \{0\}$:

[2] In fact, the results of [3] are more general, covering many families of transcendental entire functions and circle maps.

$$g_{\alpha,b}(z) = \exp(i(\theta + \alpha + b\sin(\theta))) = e^{i\alpha} \cdot z \cdot \exp\left(\frac{b}{2}\left(z - \frac{1}{z}\right)\right).$$

This function $g_{\alpha,b}$ has isolated singularities at 0 and ∞, which are *essential* (neither removable nor poles). Consequently, the behaviour near these points is very complicated; for example, the preimages of every $z \notin \{0, \infty\}$ accumulate both at 0 and at ∞. There is a set of starting values of positive area which are *escaping*; i.e., their orbits accumulate only on the essential singularities 0 and ∞.

Figure 2 illustrates the dynamics of $g_{\alpha,b}$ in the complex plane for one choice of parameters. As usual, the horizontal and vertical directions correspond to the real and imaginary parts of the starting value. This example is chaotic on the unit circle (visible in the central part of the figure), and almost every orbit is escaping. Different shades of grey distinguish different patterns with which orbits tend to the essential singularities at 0 and ∞, highlighting the intricate dynamics of the complex extension.

A crucial step in the proof of Theorem 1 is to establish *rigidity*: a map in the Arnold family cannot be deformed, by a small change of parameters, to one with the same qualitative dynamical behaviour, except through certain well-understood mechanisms. In our setting, a new difficulty arises: We must exclude the existence of deformations arising from the set of escaping points. This issue does not arise for polynomials, as treated in [2], which have no essential singularities. It is overcome by using techniques developed for studying the behaviour of transcendental entire functions near ∞ [4].

To prove the theorem, we then begin with a map $f_{\alpha,b}$ that is not hyperbolic; recall that this map has two critical points and two critical values. It follows from the above-mentioned rigidity statement that we may perturb (α, b) slightly to a parameter (α_1, b_1) for which the orbit of one critical value passes through a critical point. The set of parameters satisfying this critical relation forms an analytic variety of dimension 1. We can make another perturbation, *within this variety*, to create a

Fig. 2 The dynamics of $g_{\alpha,b}$ in the complex plane for one choice of parameters

second critical relation. (Otherwise, maps within this variety would have the same qualitative dynamical behaviour as f_{α_1,b_1}, which contradicts rigidity.)

Thus we obtain parameters (α', b'), arbitrarily close to (α, b), for which both critical values are eventually mapped to a critical point. This means that each critical point is mapped to a periodic cycle containing a critical point; such a cycle is necessarily stable. Thus f is hyperbolic, and the proof of the theorem is complete.

3 A Musical Collaboration

In 2015, Liverpool's mathematics department hosted award-winning composer Emily Howard as a Leverhulme Artist in Residence. Howard previously used mathematical and scientific ideas in her compositions, but the mathematical discussions during the residency led to the use of frontier mathematical research, rather than classical mathematical principles, in her creative process for the first time.

A particular focus of discussions during the residency were the article [3] and Theorem 1. Howard challenged the dynamics group to create sets of numeric data, encapsulating key ideas of the work. Two datasets were created by the author and Alexandre Dezotti, one of which is shown in Fig. 3. It begins by following two orbits of an unstable function in the Arnold family, starting near an unstable periodic cycle, but quickly separating due to the chaotic dynamics. As in the proof of Theorem 1, the function is slightly perturbed once, changing the parameters by less than 10^{-5}, to create an attracting cycle. A second, even smaller, perturbation yields another attracting cycle, and thus a hyperbolic function. (A member of the Arnold family with two attracting cycles is necessarily hyperbolic.) Howard's composition *Orbits* [5] is a direct creative response to this data. Other pieces (*Leviathan*, *Threnos* and *Chaos or Chess*) written during this period also used the discussions around [3] as pivotal creative input, establishing enduring principles for Howard's approach to composition.

Torus was a BBC Proms co-commission with the Royal Liverpool Philharmonic Orchestra, first performed in 2016 at the Royal Albert Hall to a sell-out audience. It was inspired by Howard's discussions with mathematicians in Liverpool, including those with the dynamical systems group, as well as conversations with Liverpool

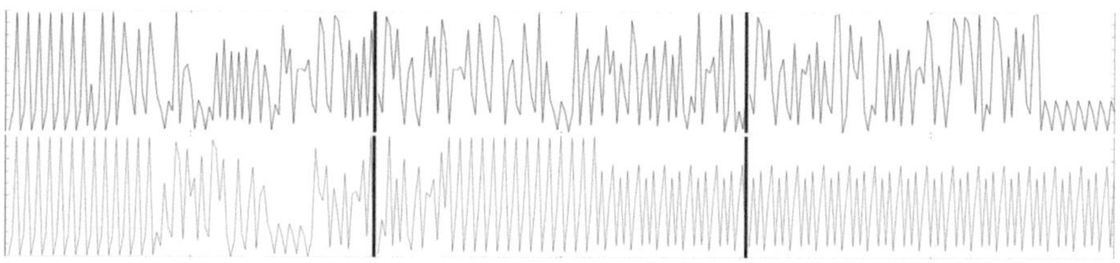

Fig. 3 The data set used in Howard's composition *Orbit 2a*

geometer Anna Pratoussevitch. The ideas of chaotic motion, and of a small perturbation that changes the fundamental nature of a system, formed an important part of Howard's compositional approach for this piece. *Torus* was hailed by the Guardian as 'one of this year's finest new works', and recognised with a 2017 British Composer Award. Two further geometry-inspired orchestral pieces followed: *Sphere* (2017) was commissioned by the Bamberg Symphony Orchestra and broadcast on BBC Radio 3, and *Antisphere* was commissioned to open the London Symphony Orchestra's 2019/20 season at the Barbican under Sir Simon Rattle.

In 2017, the Royal Northern College of Music established a new dedicated centre for Practice and Research in Science and Music (PRiSM), with Howard as Director. PRiSM is building on the approach developed during the Liverpool residency, bringing together scientists, composers and performers for mutual benefit.

Due to the specialised nature of mathematical research, it can be challenging to communicate non-superficial ideas and concepts to those outside the specific area of research, let alone outside of mathematics. There are two factors that contributed strongly to successfully creating a dialogue between music and research mathematics. We shall discuss them briefly here, as they may be instructive for collaborative projects of a similar nature.

The first is the presence of a shared language: Howard is an Oxford graduate in mathematics and computing, while the author is an amateur orchestral musician. Howard's familiarity with mathematical terminology allowed for detailed discussions of mathematical ideas, which were reflected in the resulting compositions. For instance, in the data underlying *Orbits*, there appear repeating patterns arising from both stable and unstable cycles. Though they are indistinguishable from the data alone, Howard chose to represent their differing nature musically as a result of the mathematical discussions. Likewise, the perturbations of parameters are imperceptible in the data; knowing about their significance, Howard marked them as recognisable musical events. It appears likely that, for a collaboration like this to succeed, a common language, if not already present, needs to be carefully developed.

A second key factor in the success of the collaboration was the careful choice of a specific piece of research as the basis of discussions. There are several reasons why Theorem 1 provided fertile ground for developing new creative thinking:

1. It lends itself to expression in simple and general terms: no matter how chaotic the system, there is always stability nearby.
2. Phase-locking is known and relevant to musicians, e.g. through the synchronisation of linked metronomes or the elimination of non-harmonic overtones.
3. Aspects of the proof are philosophically intriguing: the complex plane–invisible in the formulation of the problem–plays a crucial part in the proof.
4. The systems in question can be used to design data series as input into and inspiration for the creative process.

Acknowledgements The research in [3] was supported by Rempe's EPSRC Grants EP/E017886/1 and EP/E052851/1, Rempe's Philip Leverhulme Prize and van Strien's ERC Advanced Grant. Howard's residency was funded by a Leverhulme Artist in Residence award.

References

1. Arnold, V.I., *Small denominators. I. Mapping the circumference onto itself*, Izv. Akad. Nauk SSSR Ser. Mat., **25**, no. 1 (1961), 21–86.
2. Kozlovski, O., Shen, W., and van Strien, S., *Density of hyperbolicity in dimension one*, Ann. of Math. (2) **166**, no. 1 (2007), 145–182.
3. Rempe, L. and van Strien, S., *Density of hyperbolicity for classes of real transcendental entire functions and circle maps*, Duke Math. J. **164**, no. 6 (2015), 1079–1137.
4. Rempe, L., *Rigidity of escaping dynamics for transcendental entire functions*, Acta Math. **203**, no. 2 (2009), 235–267.
5. *Orbits*, by Emily Howard; see https://www.liverpool.ac.uk/mathematical-sciences/artist-in-residence/projects/orbits/.

The Evaluation of Evidence in the Form of Traces of Cocaine on Banknotes

C. G. G. Aitken and A. L. Wilson

Abstract Evidence is available of a quantity of banknotes on which traces of cocaine have been found. The question is whether this evidence is associated with a person who is or is not associated with criminal activity. A procedure for the determination of the associated likelihood ratio is described when the evidential data are believed to be driven by an underlying latent Markov chain.

1 Introduction

In a criminal case, evidence E is often evaluated through use of a likelihood ratio (LR), known as the value of the evidence, the factor which converts the prior odds in favour of a proposition proposed by the prosecution, compared with a proposition proposed by the defence, to the posterior odds. The word 'odds' is used advisedly; the propositions do not need to be exhaustive. A statistical approach is of particular interest when the evidence is in the form of data. There are many applications where such data are multivariate such as elemental compositions of glass fragments (see [1] for examples). In the example described here the data are univariate and auto-correlated. They are the quantity of cocaine on each note in a sample of banknotes as measured by a mass spectrometer. The quantity of cocaine on a note in a sample of banknotes is associated with the quantity of cocaine on adjacent notes in that sample.

The role of LRs for the analysis of data of a hierarchical random-effects nature in forensic science was introduced by [2]. One advantage of using an LR instead of a hypothesis test is that it has no dependence on an arbitrary cut-off point (e.g. 5% significance). Also an approach that is based on the LR ensures equality of

C. G. G. Aitken (✉) · A. L. Wilson
School of Mathematics and Maxwell Institute for Mathematical Sciences, The University of Edinburgh, James Clerk Maxwell Building, Peter Guthrie Tait Road, Edinburgh EH9 3FD, UK
e-mail: cgga@ed.ac.uk

A. L. Wilson
e-mail: amy.l.wilson@ed.ac.uk

P. J. Aston (ed.), *More UK Success Stories in Industrial Mathematics*,
Mathematics in Industry 42, https://doi.org/10.1007/978-3-031-48683-8_34

treatment of both propositions in contrast with hypothesis testing. The merits of the LR approach are described in [1].

The banknotes on which the quantities of cocaine were measured to determine the value of the evidence (LR) were provided by law enforcement agencies to an analytical chemistry laboratory. The laboratory measured the quantity of cocaine either on all of the notes or on a subset of the notes and is interested in the support that is provided by the results for one or other of the following propositions:

- H_C: the banknotes are associated with a person who is associated with a criminal activity involving cocaine;
- H_B: the banknotes are associated with a person who is not associated with a criminal activity involving cocaine.

Banknotes are analysed one at a time, using a mass spectrometer. The output for a sample of notes consists of a series of peaks, with each peak corresponding to one banknote. The logarithms of the peak areas (as calculated with the aid of a peak detection algorithm), which are a measure of the amount of cocaine on each banknote, are used as the data for analysis. Logarithms of the peak areas are used to reduce skewness.

Previous methods for assessment of the relative support for these propositions were concerned with the percentage of notes contaminated or else assumed independence of measurements of quantities between adjacent notes. A new method for the evaluation of this support was developed, using the LR, which accounted for autocorrelation in the data caused by the transfer of cocaine between adjacent notes and also modelled differences in contamination between different bundles of notes [3, 4].

The data that are used for the analysis, $\mathbf{z} = \{z_1, z_2, \ldots, z_n\}$ (known as the *seized sample*), are the logarithms of the peak areas corresponding to cocaine on a set of n notes of unknown association. The value of the evidence of $\mathbf{z}$ in support of H_C or H_B is to be assessed.

Training data, $\mathbf{y}$, are available, first (C), from notes from cases that went to trial and in which the defendant was convicted of a crime involving cocaine. A single criminal case may consist of multiple samples of notes, which are known as *exhibits*. There were 29 cases containing at least one exhibit with more than 20 banknotes. The 29 cases consist of between one and six exhibits. In total, there are 70 exhibits which are known to have been associated with a person who has been involved in drug crime relating to cocaine.

Training data, $\mathbf{x}$, are available, second (B), from notes that are associated with general or background circulation. There were 193 samples of notes of English or Scottish currency obtained from a variety of locations around the UK with a large number of these samples taken from the Bristol area with other samples being selected from around England, Wales and Scotland. Any set of notes from B that is used in the analyses discussed will be known as a *sample*, in contrast to an exhibit, which is used to describe notes in C.

These training data are used to develop models that are associated with H_C and H_B, respectively. The LR that is associated with the propositions H_C and H_B is given by

$$\text{LR} = \frac{f(\mathbf{z} \mid H_C)}{f(\mathbf{z} \mid H_B)},$$

where the function f is a probability density function for the measurements, conditional on H_C and on H_B in the numerator and the denominator respectively. If the LR is greater than 1, the evidence assigns more support to the proposition that the notes in the seized sample are associated with a person who is associated with drug crime involving cocaine relative to the proposition that the notes are associated with a person who is not associated with drug crime involving cocaine. The data that are analysed in [4] are available at

http://wileyonlinelibrary.com/journal/rss-datasets.

There was concern expressed in the courts (e.g., *R. v Rashid and others*, Sheffield Crown Court, January 2015) about the appropriateness of the company's database for its analyses. The database consisted mainly of notes in general circulation taken from a bank in Bristol and held, perhaps counter-intuitively, to be representative of notes in general circulation throughout Britain. In response to the courts' concern, research by [5] showed that there is no meaningful difference in quantities of cocaine on banknotes in different regions of Great Britain and hence there is no need to tailor the datasets to the region of the crime.

2 Hidden Markov Models

A hidden Markov model (HMM) was used to model the measurements of cocaine on the banknotes. In a hidden Markov model, each observed data point is associated with an unobserved state. The states form a Markov chain and determine the probability density functions of each data point conditional on the previous data point. A Markov switching model is used here, which also allows for dependence between adjacent data points, so that auto-correlation between the peak areas of adjacent banknotes is modelled.

A general sample or exhibit $\mathbf{w} = \{w_1, \ldots, w_n\}$, with n notes is considered. The states that are associated with $\mathbf{w}$ are given by $\mathbf{S} = (S_1, \ldots, S_n)$. Each state can take one of four possible values as listed in Table 1 depending on the contamination levels of the previous note and the current note. The model is shown in the Bayesian

Table 1 Hidden states in the Markov model

State S	Contamination level of previous note	Contamination level of current note
1	Low	Low
2	Low	High
3	High	Low
4	High	High

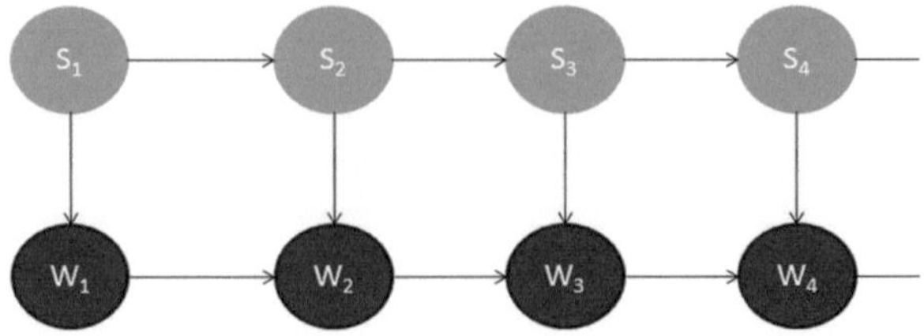

Fig. 1 Bayesian network for the HMM: nodes W denote measurements on banknotes and nodes S denote states of banknotes; subscripts denote locations of the banknote in the exhibit or sample presented to the mass spectrometer

network in Fig. 1. The observations w_i are not conditionally independent given the latent variables **S**.

Exhibits of notes tend to contain multiple bundles of notes, each of which may have a different origin, and hence these bundles may have very different levels of contamination. Samples of notes from general circulation may contain some highly contaminated banknotes which have been in contact with traces of cocaine. The states in the HMM are used to model these different levels of contamination (either high or low). The logarithm of the peak area associated with each note is modelled with a probability density function with one of four possible sets of parameter values. The particular set of parameters that is used is determined by the state of that note. The latent Markov property is required since banknotes within the same bundle (which is often secured with a rubber band or a paper strip) are adjacent to one another in the exhibit. Hence, the level of contamination of a note depends on the level of contamination of the previous note in the exhibit. The transition matrix, giving the probabilities of moving between these states, is

$$\begin{pmatrix} 1-p_{01} & p_{01} & 0 & 0 \\ 0 & 0 & p_{10} & 1-p_{10} \\ 1-p_{01} & p_{01} & 0 & 0 \\ 0 & 0 & p_{10} & 1-p_{10} \end{pmatrix}.$$

Normal, beta, gamma and inverse gamma prior distributions are chosen as appropriate. The training data are used in combination with prior distributions to estimate the following parameters, using a Metropolis-Hastings sampler.

- Means and variances for the two states (high and low) for the quantity of contamination.
- The autocorrelation coefficient for lag 1.
- The transition probabilities p_{01} and p_{10}.
- Hyperparameters for the prior distributions for the two variances.

The initial distribution of the four hidden states was taken to be fixed and known, with each probability equal to 0.25. Further details are given in [4].

3 Results

The performance of the model was assessed with the rate of misleading evidence. Each exhibit and sample was taken out in turn and treated as the seized sample of banknotes, **z**. The other samples and the exhibits from other criminal cases (i.e. removing the chosen exhibit or sample and any exhibits from the same criminal case as the chosen exhibit if the seized sample were known to be from the set C) were then treated as the measurements **x**, and **y**, respectively, and used to calculate the LR for the sample seized. If the LR for the seized sample was greater than 1, **z** was said to provide support for proposition H_C and, if it was less then 1, **z** was said to provide support for proposition H_B. The proportion of times that the seized sample provided support for the proposition with which it was not associated is given as the rate of misleading evidence.

For exhibits, the rate of misleading evidence was 35.7% (25/70), for general circulation samples it was 10.4% (20/193). The large rate of misleading evidence for exhibits is expected as many of the exhibits were not contaminated. However, these rates do not give an indication of the size of the LRs that were obtained. An LR that is close to 1 does not provide strong support for either proposition. A method which gives larger LRs, where these large values are justified, is of more value when being used in the evaluation of evidence. A Tippett plot considers the actual values of the likelihoods and the LRs for the HMM. Figure 2 indicates the proportion of general circulation samples (broken curves) and crime exhibits (full curves) that have a log-likelihood ratio greater than the value on the horizontal axis of the plot.

The HMM allows for two different levels of contamination within samples and exhibits, as well as accounting for auto-correlation between measurements on adjacent banknotes. Failure to account for auto-correlation, when it is known to be present, creates a risk that LRs will be overstated. Failure to account for different levels of contamination on different bundles within samples and exhibits, means there is potential for understating LRs.

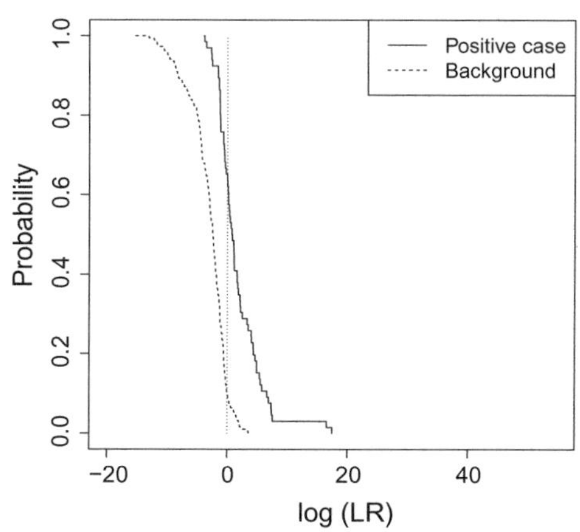

Fig. 2 Tippett plot of HMM (solid line, exhibits; dashed line, samples)

4 Impact

The Scientific Director of MSA commented that '[t]he work . . . is routinely referred to in the supporting material [in] every court statement sent out by [MSA]. . . As such, it is frequently subject to cross-examination in Court, and the expert witnesses make reference to the peer-reviewed papers regularly. It is difficult to quantify, but the experience of the court attending witnesses is that ensuring that our offering is on a sound scientific footing has greatly reduced the "attacks" on our methodology.' and 'the work [5] demonstrating that the general circulation background samples do not vary greatly across the country has proved to be enormously useful following adverse criticism of the sampling strategy in the trial of *Rashid and others* in Sheffield Crown Court in 2015. This work is referred to in talks given to police officers for marketing purposes, and is included in the "In-house method validation" documentation provided to the United Kingdom Accreditation Service at Accreditation Assessment visits.'

More broadly, the research in [3, 4] has changed the way in which forensic scientists at MSA think about the presentation of their forensic evidence, resulting in improvements in the administration of justice. MSA has provided evidence for around 200 cases per year in the UK (most of the UK cases featuring this evidence type). Further to the above, Aitken has given evidence as an expert witness based on research in [5] under oath in two trials: *R. v. Hussain and others* (Snaresbrook Crown Court) and *R. v. Parry and others* (Liverpool Crown Court).

Acknowledgements The research described here was supported by an EPSRC 'Cooperative award in science and engineering' (CASE), award (009002219), sponsored by Mass Spec Analytical (MSA) of Bristol who provided the data and much valued assistance. Key Forensic Services purchased the banknotes part of MSA in 2020 and continued using the research described. The authors are particularly grateful to Richard Sleeman, the Scientific Director of MSA at the time of the research, who died in September 2023. His vision and support were crucial for the success of this work.

References

1. C G G Aitken, F Taroni, and S Bozza. *Statistics and the evaluation of evidence for forensic scientists*. John Wiley and Sons Ltd., Chichester, third edition, 2021.
2. D V Lindley. A problem in forensic science. *Biometrika*, 64:207–213, 1977.
3. A Wilson, C G G Aitken, R Sleeman, and J Carter. The evaluation of evidence relating to traces of cocaine on banknotes. *Forensic Science International*, 236:67–76, 2014.
4. A Wilson, C G G Aitken, R Sleeman, and J Carter. The evaluation of evidence for auto-correlated data in relation to traces of cocaine on banknotes. *Applied Statistics*, 64:275–298, 2015.
5. C G G Aitken, A Wilson, R Sleeman, B Morgan, and J Huish. Distribution of cocaine on banknotes in general circulation in England and Wales. *Forensic Science International*, 270:261–266, 2017.

Maintaining Standards: Metrology and Statistics Combining to Produce the Primary Reference Materials Which Underpin All Radiocarbon Dating Measurements

E. Marian Scott

Abstract Underpinning radiocarbon dating is a series of natural reference materials that have been created as a result of statistical research in first the design of proficiency trials and then secondly the analysis of the resulting data. This short article provides an overview of the statistical contributions to this field, which provides the chronological basis of many other research disciplines including but not limited to Archaeology.

1 Introduction

The long half-life of radiocarbon of 5,730 years has proved revolutionary in cultural and environmental heritage activities, such as archaeology and forensic science by facilitating measurement of the age of an artefact based on its radiocarbon content. The measurement is based on a complex process involving pretreatment of the sample (which typically is wood, bone, shell, peat to mention a few), followed by transformation to, most commonly, a graphite target. The resulting unknown age graphite target is then measured in an accelerator mass spectrometer (AMS) along with a series of graphite targets from known age materials, standards and background samples. In such a complex measurement process, quality assurance and quality control are vital aspects to ensure the accuracy and precision of any measured radiocarbon date. Like every complex measurement process, each measurement has an uncertainty, sometimes described as its error, which fundamentally is a quantification of the variability that would be observed were we able to make true repeated measurements. Contributions to this uncertainty come from measurement of standards (of known activity), backgrounds (no ^{14}C activity), known activity reference materials (which allows the evaluation of bias which is sometimes included in the definition of uncertainty (described as being systematic)) and other sources, which potentially include

E. Marian Scott (✉)
School of Mathematics and Statistics, University of Glasgow, Glasgow G12 8QW, UK
e-mail: Marian.Scott@glasgow.ac.uk

P. J. Aston (ed.), *More UK Success Stories in Industrial Mathematics*,
Mathematics in Industry 42, https://doi.org/10.1007/978-3-031-48683-8_23

technician effects, pre-treatment effects as well as other contributions not easily uniquely identified, such as the intrinsic variability within natural materials. There are approximately 100 radiocarbon laboratories worldwide, and more than 50,000 radiocarbon determinations made annually. The reputation of this global community of laboratories is dependent on the use of some unique certified reference materials developed as a result of our research which has included 7 global proficiency trials, as well as other quality assurance procedures. Our quality assurance research programme provides underpinning certified reference materials which play a key part in ensuring measurement consistency and quality across the global laboratory and user communities. Figures 1 and 2 show some of the preparation and final samples used for the most recent radiocarbon intercomparison.

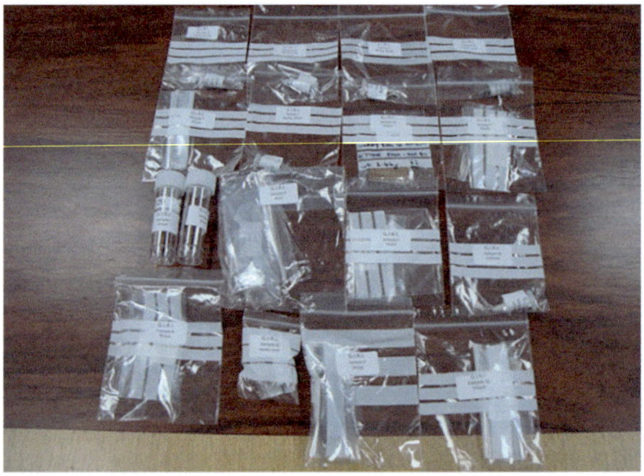

Fig. 1 The most recent intercomparison samples prepared for each lab

Fig. 2 Individual tree-rings being extracted from a block

2 Underpinning Statistical Research

Key to any quality assurance programme is the design of the experiment that the laboratories will take part in. This can be considered primarily as an experiment to quantify variation and accuracy, with the most recognised form of experiment being an inter-comparison exercise, sometimes also called a round-robin or proficiency trial. The stages in designing such an experiment include identifying the samples that will be used (they should be homogeneous in C-14, available in sufficient quantity, and span the radiocarbon dating range (from modern to more than 50,000 years). At the same time, the statistical design principles of blindness of the laboratories to the age of the samples, the use of duplicate or replicate samples for direct estimation of laboratory uncertainty are followed. Finally, the samples should test the different stages of the dating process, so a hierarchical design is used. The principles that we followed in the creation of the trial designs we have used are:

- A series of unrelated individual samples, spanning the dating age range
- Some linked samples to earlier intercomparisons to allow traceability
- Some known age samples, to allow independent accuracy checks
- A small number of duplicates, to allow independent estimation of laboratory uncertainty.

Our statistical research has been key in creating the radiocarbon reference materials, including identifying plausible natural materials for testing, evaluating sample homogeneity, preparing and designing the testing protocol through a series of statistically designed, hierarchical inter-comparisons (four global inter-comparisons since 2000 [1]) involving laboratories based in more than 50 countries with each inter-comparison involving between 60 and 75 laboratories. On receipt of the radiocarbon results from the participating laboratories, different statistical models including hierarchical random effects models were used to characterise each material (i.e. define the agreed/certified/consensus known age) and quantify variation. These models take into account laboratory estimated variation and biases before estimation of the consensus age and associated uncertainty.

3 Statistical Models

The basic measurement model assumes that a ^{14}C age determination X_i is subject to variability due to some known and potentially unknown laboratory and environmental sources, and that this is quantified in the so-called counting error, which is an uncertainty quantification unique to the individual age determination. It is perhaps simplest to consider the counting error as the culmination or propagation of a series of contributing uncertainties or sources of variation, forming a hierarchy. Focussing on the AMS context, measurements made in the same batch (AMS wheel) may share common error components (e.g. same standard or backgrounds being used within an AMS target wheel, same operator, same operating mode). An important practical and philosophical point is, if the determination X_i were repeated many times (say n), with all conditions apparently being kept absolutely constant, then there would be a distribution of results $X_i \sim N(\mu, \sigma^2)$, typically assumed a Normal distribution. It would then be possible to estimate the dispersion, σ^2 in the distribution of the set of measurements by the sample variance, providing an empirical quantification of the measurement precision. The interesting and relevant question is how the sample variance in such a set relates to the typical quoted error values. As well of course, we are not able to make absolute replicate measurements since we cannot control all sources of variability within the sample or the laboratory or the machine.

The resulting data from the proficiency trial can be considered as follows: for each sample, we have a series of measurements, some are independent, some may be dependent (more than one measurement from the same laboratory).

So having designed the experiment (or laboratory intercomparison), how do we analyse the results? The first step in the process is to estimate the consensus value for each sample. The description below summarises an algorithm which has been used in many studies [2].

- **Stage 1**: Outlying results are removed if they are greater than 3 $\times$ inter-quartile ranges from the nearest of either the lower or upper quartiles. The preliminary consensus value is calculated as the median (m) of the remaining results.
- **Stage 2**: Remove results that are at least twice their quoted error(s) from the preliminary consensus value. That is, only keep $|x - m|/s < 2$, where x is the result, m the preliminary consensus value and s the quoted error. The rationale for this step concerns the relationship of the result and its quoted error. Results that lay a long way from the preliminary consensus value in relative terms should be down-weighted.
- **Stage 3**: Calculate the final consensus value as a weighted mean of the remaining results and of course generate the appropriate uncertainty of that value.

More recently, we have also adopted a second model using a linear mixed effect model that reflects the more common case where AMS facilities are likely to report several determinations for the same sample. This allows the common sources of variation to be correctly attributed. To estimate the consensus value for each material, using the linear mixed (or random effects) model, the total variation in F (or age) around the 'true' age for the material is attributed to two components, the within laboratory (common source of variation), and the between laboratory variation.

With the consensus values then calculated for each sample, we can then turn our attention to evaluating laboratory performance, this includes evaluating a laboratory bias or offset term as the average deviation (weighted) from the consensus value. Importantly we have also been able to extend this model to include an error multiplier (so a multiplicative error model rather than additive one) and which allows a further investigation of the appropriateness of the quoted errors.

The simplest measurement model in this case is that

$$X_{ij} \sim N(\mu_i, \theta_j \sigma_{ij}^2)$$

where X_{ij} is the ^{14}C age for sample i as measured in lab j, μ_i is the true age, σ_{ij} is the quoted error for sample i and $\sqrt{\theta_j}$ is the error multiplier for laboratory j. This model is then further developed to also allow the evaluation of a laboratory offset, α_j. In the simplest form, θ_j is a multiplicative error term, which is straightforward to estimate.

Finally, we have made use of z-scores [3] to evaluate overall performance. Interpretation of z-scores includes accuracy, precision and 'fitness for purpose'.

For the analysis, we have reported z-scores for a laboratory/material calculated as

$$Z_{ij} = (X_{mij} - X_{ai})/s_{pij}$$

where X_m, is the reported result for the material, X_a is the assigned or true value for the material, s_p is the target value for the standard deviation for values of X, i represents material and j represents laboratory. We have used the laboratory quoted error for s_p though hypothetically it is determined by fitness for purpose and represents the amount of uncertainty in the results that is tolerable.

Interpretation of the z-score reflects the accuracy achieved and provides a means of making a judgement concerning fitness for purpose.

It is commonly assumed that z should be normally distributed with zero mean and variance 1, where:

- A z-score of 0 implies a perfect result
- A z-score between -2 and $+2$ is generally considered as complying with fitness for purpose
- A z-score outside -3 or $+3$ would be very unusual and further investigation would be needed

We can also consider how to modify this basic formulation to account as well for the uncertainty that is associated with the measured X.

4 Details of the Impact

Using the set of all intercomparisons results, based on random effects models, we have been able to estimate uncertainty as a function of material and age ($> 45,000$ years to modern). Each past inter-comparison, the fourth (FIRI, 1997–2002), fifth (VIRI, 2004-2008) and sixth (SIRI, 2013–2017), has expanded the repository of reference materials that are available for laboratories (on request) for quality assurance purposes [1]. During the calibration curve estimation (IntCal20 [4]), prior specification of sources of uncertainty are required (based on replicate measurements) and results from the inter-comparisons have contributed to those prior specifications.

The statistical research has been fundamental to the development of metrological principles and good practice for Radiocarbon laboratories and users. We have led the way in the development of statistical models for the complex data generated in laboratory inter-comparisons, and specifically how the material consensus value is evaluated with its associated uncertainty. These approaches have since been adopted in other cosmogenic isotope communities, including beryllium and aluminium, which are key in geological dating and the basic statistical model was first used by the International Atomic Energy Agency (IAEA) and continues to be used by them in characterising their set of radiocarbon standards. University of Glasgow statistical research has demonstrated several key findings for end users such as archaeologists (in national museums, in rescue excavations, in cultural heritage), geologists, the antiquities and nuclear industry sectors. Findings included:

- The evidence for significant laboratory biases and their magnitude
- That laboratory quoted errors are frequently underestimated, and how they should be adjusted to be realistic using error multipliers; these first two findings have allowed laboratories to explore their procedures and to identify any corrective steps needed
- The statistical quantification of uncertainty, intrinsic in natural materials and laboratory procedures [5]. This has allowed proper and full quantification of uncertainty, to be incorporated into calibration procedures and ultimately ensuring that chronologies are formed and used with realistic uncertainties

Acknowledgements Thanks to all the sample providers and to the laboratories who have participated.

References

1. Scott, E.M., Naysmith, P., Cook, G.T.: Why do we need ^{14}C inter-comparisons?: The Glasgow ^{14}C inter-comparison series, a reflection over 30 years. Quat. Geochronol. **43**, 72–82 (2018)
2. Scott, E.M., Cook, G.T., Naysmith, P.: Error and uncertainty in radiocarbon measurements. Radiocarbon **49**, 427–440 (2007)
3. Thompson, M., Ellison, S.L.R., Wood, R.: The international harmonized protocol for the proficiency testing of analytical chemistry laboratories (IUPAC technical report). Pure Appl. Chem. **78**, 145–196 (2006)
4. Heaton, T.J., Blaauw, M., Blackwell, P.G., Bronk Ramsey, C., Reimer, P.J., Scott, E.M.: The INTCal20 approach to radiocarbon calibration curve construction: A new methodology using Bayesian splines and errors-in-variables. Radiocarbon **62**, 821–863 (2020)
5. Scott, E.M. (ed.): The third international radiocarbon inter-comparison (TIRI) and the fourth international radiocarbon inter-comparison (FIRI) 1990–2002: results, analyses, and conclusions. Radiocarbon **45**, 135–408 (2003)

Improving Population Censuses Around the World

Paul A. Smith, Peter G. M. van der Heijden, and Li-Chun Zhang

Abstract Population censuses provide a wide range of detailed information, but are expensive and infrequent. Some more frequent information can be obtained using administrative data sources, which can also be combined with some directly collected information. This combination of sources requires new approaches to produce objective census estimates. We apply methods to deal with undercoverage and overcoverage in administrative data in different contexts, and to examine the quality of population estimates based on administrative data sources, based on the application of dual and multiple system estimation. We also apply novel design methods to the traditional census operation in the UK.

1 Introduction

The periodic population census is a crucial component of a nation's statistical system, providing an opportunity to produce detailed statistical estimates nationally and for small domains (both geographical areas and subpopulations). These statistics have many uses, including for local planning and monitoring, as a component of population estimates and social statistics, and for formation and evaluation of government policy. They have a long shelf-life because the census is expensive (e.g. the England & Wales

P. A. Smith (✉) · P. G. M. van der Heijden · L.-C. Zhang
University of Southampton, Highfield, Southampton SO17 1BJ, UK
e-mail: p.a.smith@soton.ac.uk

P. G. M. van der Heijden
e-mail: P.G.M.vanderHeijden@uu.nl

L.-C. Zhang
e-mail: L.Zhang@soton.ac.uk

P. G. M. van der Heijden
Utrecht University, Utrecht, Netherlands

L.-C. Zhang
Statistics Norway, Oslo, Norway

P. J. Aston (ed.), *More UK Success Stories in Industrial Mathematics*,
Mathematics in Industry 42, https://doi.org/10.1007/978-3-031-48683-8_36

census in 2021 cost around £900 m) and only undertaken periodically. Traditional census enumeration involves asking every household to complete a questionnaire, but some countries, including the Netherlands, use a census based on administrative data. This reduces field costs and potentially improves the frequency of estimates, and many more countries are considering similar approaches.

The main challenge in a traditional or administrative data census is to estimate population sizes (overall and in subgroups) using imperfect data. Traditional census enumeration suffers from selective non-response, and it is often groups of policy interest which are under-represented. Tailored designs are therefore needed so that relevant data can be collected and used with estimation approaches to generate high quality estimates (Sect. 2). This is traditionally done with dual system (capture-recapture) estimation, based on the census operation and a post-enumeration survey (also known as a coverage survey) as the two sources. The inclusion or not of records in these two data sources form a contingency table, with the records not in either source unobserved. The entries in this table can be modelled with a loglinear model, and the parameters used to estimate the number of unobserved people. A similar approach can be used with administrative data sources, but these are prone to both under- and over-coverage, and the presence of overcoverage violates one of the assumptions for dual system estimation. Statistical approaches for estimating population sizes using sources with overcoverage and with inconsistent characteristics between sources are considered in Sects. 3 and 4.

2 Sample Design for a Traditional Census Operation

In 2001 the UK[1] developed the strategy of a population census and a coverage survey (or post-enumeration survey) to explicitly make population estimates based on dual system estimation. These estimates in areas sampled in the coverage survey were then used as inputs to a model using the observed census count as a predictor variable, in order to make estimates for all areas. These were then used to produce a fully adjusted database [see 1]. The design of the coverage survey in 2001 was conservative, as there was little prior information on which to base any more detailed assessment, but the 2001 census provided substantial information to use in the design of the 2011 coverage survey.

There were several design features to be optimised. The census coverage survey (CCS) was stratified by geography, then by a Hard to Count (HtC) index which acted as an indicator for the expected response rate. The index had to be constructed and different stratification evaluated against a design variable. Creating a suitable design variable is challenging–it has to summarise 37 age-sex variables, and two design variables were calculated as residuals from a simple model, one each for persons and households. These were used to allocate sample to different types of

[1] The population censuses in the UK are run separately, but in coordinated fashion in England & Wales, Scotland and Northern Ireland.

areas, and demonstrated that stratification by Local Authority was better as a basis for estimation. A similar investigation showed that a finer HtC stratification, with separate strata for the most challenging areas, improved the census estimates.

The CCS was clustered to make the fieldwork efficient, but by choosing more, smaller clusters the design can be made more efficient (with smaller sampling variance). This, however, has an effect on costs, so in order to evaluate the options a cost model was developed for the CCS fieldwork operation. This demonstrated that a move to Output Areas as the cluster would be more efficient with little change in the fieldwork costs.

Several other features and options for the design of the CCS were investigated to optimise the design, while still leaving the design general enough to work well with the (unknown) changes in 2011. For more details see [2].

3 Population Census Based on Administrative Data Sources

The high cost of traditional census operations has led to strong interest in alternative approaches. Some countries have person registration systems, which provide a sound basis for an administrative data census. The Netherlands operates such an administrative data census called the *virtual census*. Although the administrative sources on which it is based are good quality, there is still some undercoverage, and it is of particular interest to estimate the undercoverage of specific populations. Bakker et al. [3] use dual system estimation (DSE) with the person register and the police register, which records details of suspects apprehended by the police. By matching people in these two sources, an estimate of the unregistered population can be obtained. However, additional detail of the characteristics of this population are needed, and further variables such as sex can be used to refine these estimates. Additional categorical covariates can be included in the model, and [3] also uses ethnicity to identify undercoverage of people from diverse clusters of countries, for example the Afghan, Iraqui and Iranian people, not included in the population register.

A range of countries including the UK, Ireland, Italy and New Zealand are all seeking to move from periodic large-scale censuses to a virtual census similar to the Netherlands model, based on administrative datasets. This has significant advantages in reduced and (crucially) more evenly spread costs, and also in the potential to update census estimates much more frequently. For countries that do not have a person register, such as the UK, Ireland and New Zealand, one must rely on linked administrative data sources only, which can suffer considerable overcoverage in addition to undercoverage. Overcoverage in the lists used in DSE (for undercoverage adjustment) translates directly to bias in the population size estimates, so we would like to remove it. This cannot be done directly in estimation without an independent estimate of the overcoverage.

Zhang and Dunne [4] introduce a strategy for dealing with overcoverage in administrative data sources, called the *trimmed DSE*. The basic idea is to successively remove records at risk of overcoverage from the administrative data, calculating

DSE at each step. When the sequence of DSE's stops declining, this is an indicator that the overcoverage has been removed and that only undercoverage is left. The DSE at this point meets the required assumptions. This provides an objective way to remove the overcoverage and provide a valid population estimate.

The trimmed DSE has already been implemented by the Central Statistics Office in Ireland as part of the evaluation of the quality of administrative sources for use in an administrative data census. As Ireland does not have a population register, the basic idea is to let the driver license registrations (new and renewal) play the part of a coverage survey while other selected administrative sources are linked and processed to play the part of an (administrative data) census that suffers only from undercoverage. The trimmed DSE is helpful for de-selecting any source that is deemed unsuitable due to the overcoverage it would bring to the system, as well as assessing whether overcoverage remains notable in the integrated administrative file afterwards. For more details see [4] and [5].

4 Undercoverage and Inconsistency in Linked Administrative Data Sources

In New Zealand there is also interest in developing a population census based on administrative data, and there are three administrative data sources with wide coverage which can be used for this purpose, as well as the five-yearly population census. They are the birth register, an education register and the health register. Here again we are interested in estimating the size of the population using multiple system estimation (MSE), by estimating how many people are missed by all the lists. We can consider a situation where the administrative sources are used alongside the census, or where only the administrative sources are used. MSE is quite well-known, but here there is an additional challenge that the birth and education registers have only partial coverage of the population. This raises the question of whether the standard estimator will produce unbiased estimates. This has previously been considered for two lists by [6], but here we needed a more general approach.

A range of models can be fitted to the observed numbers, depending on which interactions between lists are included. These models fall into groups according to the estimated number of observations not in any list–different models produce the same estimates. This leads to a theory of *invariant population size estimation* which explains under what conditions we expect the estimates to be the same [for a review see 7]. We can use the same arguments to explain when partial coverage in some of the sources can be 'ignored' because it will not affect the estimate of the size of the population not observed in any list [see 8]. It turns out that the partial coverage of two of the four lists in the New Zealand case can be ignored because of their relation to other parts of the data, and therefore the modelling proceeds as if these lists covered the whole population.

For an administrative data census, estimating the general population size is only part of the challenge; to replace the traditional census it also needs to provide estimates of the numbers of people with certain characteristics. In the New Zealand situation we consider ethnicity as an example, initially aiming to estimate the population size broken down into Māori and non-Māori categories. The population census and the three administrative sources all collect ethnicity information, but people are not always consistent in their answers to the ethnicity questions, and in some cases refuse to answer so that ethnicity is missing. This leaves three issues to solve–estimating the population size, imputing an ethnicity in the cases that it is not observed, and resolving the inconsistencies where different ethnicities are provided.

We can treat this data as an 8-way contingency table, four margins for inclusion or not in the data sources (with the cell for not included in any source unobserved), and four margins for the ethnicity (Māori/ non-Māori). We use the EM algorithm to fit a loglinear model and use the parameters to split out both the missing ethnicity observations and the estimate for the unobserved part of the population between ethnicity categories. The high coverage and large numbers of people mean that we can use a model with many interactions, but the maximal model is unstable, and some detailed interaction terms need to be dropped. The estimation process uses the interaction patterns in the observed part of the data to impute in the unobserved parts. This gives an estimate for the population size by ethnicity, but still produces different ethnicity estimates in each source–that is, the ethnicity variables measure a slightly different concept in each data source.

In order to have a consistent ethnicity estimate, we fitted a latent class model, initially to the outcome of the estimation process previously described. Then we argued that including the latent class model within the model fitted in the EM algorithm is better because it accounts for all the variability in the data in one fitting procedure. Some interactions which are now explained by the relation to the latent variable need to be removed so that the model is consistent with the assumptions of latent class models. As expected, the two latent classes that are estimated are consistent with the two ethnicity labels that are present in the original data, and this allows us to interpret the measurement errors of ethnicities in the different data sources. The population census has been generally held to be the best measure of Māori ethnicity among the different sources available in New Zealand, and it has low values for measurement error in both Māori and non-Māori. The Health register has a low error for non-Māori, but the largest measurement error for Māori. The birth and education registers are similar to the census in the estimated measurement error in the Māori class, but have more error in estimating the non-Māori class. Therefore overall our results support the conclusion that the census is the best overall measure of ethnicity. For more details see [8].

5 Conclusion

The design work on the traditional census operation demonstrated that the variance of 2011 England & Wales census estimates could be reduced for the same cost, and this was implemented, leading to high quality outputs based on dual system estimation from the census enumeration and a coverage survey. The trimmed DSE and the MSE approaches provide a principled approach to constructing a census based wholly or partly on administrative data sources. The trimmed DSE is a way to deal with overcoverage, which is particularly valuable when there is no person register, as in Ireland. The latent class multiple system extends the approach to deal with inconsistent information in administrative data sources. Both approaches facilitate a census based on administrative data, which is cheaper and easier to manage than a periodic large-scale census operation, and can produce estimates more frequently.

Acknowledgements We are grateful to the staff (too many to name individually) of the Office for National Statistics (UK), Central Statistics Office (Ireland), Statistics Netherlands and Statistics New Zealand for collaborating in the development and application of the methods described here.

References

1. James J Brown, Christine Sexton, Owen Abbott, and Paul A Smith. The framework for estimating coverage in the 2011 census of England and Wales: combining dual-system estimation with ratio estimation. *Statistical Journal of the IAOS*, 35(3):481–499, 2019.
2. James Brown, Owen Abbott, and Paul A Smith. Design of the 2001 and 2011 census coverage surveys for England and Wales. *Journal of the Royal Statistical Society: Series A (Statistics in Society)*, 174(4):881–906, 2011.
3. Bart FM Bakker, Peter GM van der Heijden, and Susanna C Gerritse. Estimation of non-registered usual residents in the Netherlands. In D. Böhning, P.G.M. van der Heijden, and J. Bunge, editors, *Capture-recapture methods for the social and medical sciences*, pages 259–273. Chapman and Hall/CRC, 2017.
4. Li-Chun Zhang and John Dunne. Trimmed dual system estimation. In D. Böhning, P.G.M. van der Heijden, and J. Bunge, editors, *Capture-recapture methods for the social and medical sciences*, pages 237–257. Chapman and Hall/CRC, 2017.
5. John Dunne and Li-Chun Zhang. A system of population estimates compiled from administrative data only (with discussion). *Journal of the Royal Statistical Society: Series A (Statistics in Society)*, 187(1):3–38.
6. Eugene N Zwane, Karin van der Pal-de Bruin, and Peter GM van der Heijden. The multiple-record systems estimator when registrations refer to different but overlapping populations. *Statistics in Medicine*, 23(14):2267–2281, 2004.
7. Peter GM van der Heijden, Paul A Smith, Joe Whittaker, Maarten Cruyff, and Bart FM Bakker. Dual-and multiple-system estimation with fully and partially observed covariates. In L.-C. Zhang, & R. Chambers, editors, *Analysis of Integrated Data*, pages 137–168. Chapman and Hall/CRC, 2019.
8. Peter GM van der Heijden, Maarten Cruyff, Paul A Smith, Christine Bycroft, Patrick Graham, and Nathaniel Matheson-Dunning. Multiple system estimation using covariates having missing values and measurement error: Estimating the size of the Māori population in New Zealand. *Journal of the Royal Statistical Society: Series A (Statistics in Society)*, 185(1):156–177, 2022.

MIX
Papier aus verantwortungsvollen Quellen
Paper from responsible sources
FSC® C105338

If you have any concerns about our products,
you can contact us on
ProductSafety@springernature.com

In case Publisher is established outside the EU,
the EU authorized representative is:
Springer Nature Customer Service Center GmbH
Europaplatz 3, 69115 Heidelberg, Germany

Printed by Libri Plureos GmbH
in Hamburg, Germany